1,000,000 Books

are available to read at

---◆---

www.ForgottenBooks.com

---◆---

Read online
Download PDF
Purchase in print

ISBN 978-1-331-03186-4
PIBN 10135884

1 MONTH OF
FREE
READING

at

www.ForgottenBooks.com

By purchasing this book you are eligible for one month membership to ForgottenBooks.com, giving you unlimited access to our entire collection of over 1,000,000 titles via our web site and mobile apps.

To claim your free month visit:
www.forgottenbooks.com/free135884

English
Français
Deutsche
Italiano
Español
Português

www.forgottenbooks.com

Mythology Photography **Fiction**
Fishing Christianity **Art** Cooking
Essays Buddhism Freemasonry
Medicine **Biology** Music **Ancient
Egypt** Evolution Carpentry Physics
Dance Geology **Mathematics** Fitness
Shakespeare **Folklore** Yoga Marketing
Confidence Immortality Biographies
Poetry **Psychology** Witchcraft
Electronics Chemistry History **Law**
Accounting **Philosophy** Anthropology
Alchemy Drama Quantum Mechanics
Atheism Sexual Health **Ancient History**
Entrepreneurship Languages Sport
Paleontology Needlework Islam
Metaphysics Investment Archaeology
Parenting Statistics Criminology
Motivational

NATURAL RESOURCES

OF THE UNITED STATES

BY

JACOB HARRIS PATTON, M. A., PH. D.

*Author of "A Concise History of the American
People"; "Yorktown, 1781-1881"; "Our Tariff";
"Brief History of the Presbyterian Church in the
United States, from 1705 to 1888"; "Natural
Resources of the United States" (Primer), etc.*

PUBLISHED, 1888, BY
D. APPLETON AND COMPANY, NEW YORK
LONDON: PATERNOSTER SQUARE

PREFACE.

———

THE intention of this volume is to give the American people a concise narrative of the natural resources of their own country, in *all* their numerous forms. Only four classes of these treasures have been written upon—the precious metals, coal, iron, and petroleum. The present view is designed to be comprehensive and sufficiently full on each resource; it proposes to give an account only of the materials found in Nature, and on which are based the industry and the physical comfort of the people. "The original source of wealth is the bounty of God in Nature." It is not within the scope of this volume to give information in respect to the using of these sources of wealth, such as of the means of transportation or of manufacturing, and yet it has occasionally related incidents connected with such beginnings—the latter being a class of information that would interest the intelligent reader.

The question arises, Is it possible, in a book of 523 pages, to give a satisfactory account of these resources? In the volumes published on the precious metals, and on coal and iron, more than nine tenths of their pages are taken up in giving details of the means of transportation, and of mining companies and their operations. Such information is valuable to the practical miner or railway manager, but is of little interest or profit to the general reader. On the contrary, that these treasures are stored in the earth to confer blessings upon the people of the United States, not only at the present time, but for generations to come, is a truth recognized and appreciated by every intelligent man and woman in the land, as in this bounty of Nature they *all* have an interest.

These resources are remarkable for their vastness; but equally striking has been that providential care which provided, under such peculiar circumstances, a Christian people—lovers of liberty, civil and religious—to occupy this goodly land, and, by their energy and industry, bring into practical use these varied treasures. (*See History of the American People.*)

It is a common error, at first thought, to reckon gold and silver as the most important source of the wealth of the Nation; though they are far transcended in value by the coal and the iron, while the latter two are fully as far surpassed in worth by the fertile soil, the rainfall, and the sunshine. It is not strange that these errors exist, since the attention of the American people has never been directed to a comprehensive and a comparative view of *all* the

varied and immense resources with which the Creator has endowed their land. In this volume, full outlines are given of these treasures, in their numerous
forms, their amount and characteristics, in order
that the intelligent reader may have a definite view
of the whole.

In preparing this work the author was greatly
encouraged by the unusual interest manifested in his
effort to lay before the American people a summary
of their native wealth. For instance, *thirty-one Governors,* of the thirty-two invited, responded liberally
to the request for information in respect to the various resources of each one's State or Territory; in a
similar manner was he aided by the Interior Department at Washington. Special thanks are due to
Prof. David T. Day, of that department, in furnishing the author, in advance of their publication, a summary of the output, etc., of the year 1886: the latter
will be found in the tabulated column for that year
at the end of the volume. The Governors sent in all
ninety-four volumes, ranging in size from 1,180 pages
down to a dozen or more, in pamphlets, besides a
number of manuscript letters from officials supplying deficiencies. In addition were numerous letters
and items on the subject from other authentic sources.
Due credit is given in the body of the work to the authorities consulted.

The author hopes that, by means of the facilities
thus afforded, he has been able to put in form an account of the subject so fully as to meet the wants of

the general reader and student. He has endeavored to cover the entire field, as these resources do not increase, and only diminish as the people utilize them: in consequence, when *once described and located,* the work needs no further extension, except in case of new discoveries, when an account of such will be promptly inserted. That the reader may have a conception of the continuous progress of the Nation's development, or rather its practical use from year to year of its resources, there is prepared at the end of the volume a tabulated summary of the output of the mines, etc., for three years, so that from this table a *pro rata* estimate can be made of the Nation's future material progress. The intention is to add to this tabulated summary from year to year, as new matter or information demands. The volume has a complete index, which is so essential to the convenience of the reader or student.

Attention is occasionally directed to the immense value of the natural resources of the United States, when compared with those of other countries. TRUTH IS NOT A BRAGGART, and it is proper that the fact in this relation should be known, that the American people, especially the younger portion, may appreciate more fully the natural advantages of their own country.

J. H. P.

NEW YORK CITY, *November 15, 1887.*

CONTENTS.

Contents.

The original source of wealth is the bounty of God in Nature.

DR. FRANCIS WAYLAND.

The material prosperity of the United States is based upon the development of the resources with which they are so richly endowed.

Nature spontaneously furnishes the matter of which all commodities are made.   MCCULLOCH.

Nature does more than supply materials ; she also supplies powers.   JOHN STUART MILL.

THE NATURAL RESOURCES
OF THE UNITED STATES.

INTRODUCTION.

THE territory of the United States, exclusive of Alaska, may be characterized as a land with a sunny exposure, as the belt of the continent thus occupied, and which extends from the Atlantic to the Pacific, greets the sun by facing the south. Under laws instituted by the Creator, and which govern the air when in motion, the winds near the surface of the earth, being colder and heavier, when not interrupted, flow from the north toward the south, where, becoming warmer, and in consequence lighter, they float upward, thus producing a partial vacuum, and keep continuously restoring the equilibrium by returning, in the higher regions of the atmosphere, from the south toward the north. This circulation of the air secures to the people of the United States the blessings of a climate having, it is true, a varied temperature, but graduated in such manner as to invigorate the human constitution. No such movement of the winds could occur uniformly in an east and west direction round the globe, since on the same parallels of latitude the temperature is always nearly the same, and, by the laws of Nature just noted, there could never exist to any extent a partial vacuum or an equilibrium to be restored; in consequence, the variations in the winds—east and west—that may happen on any parallel, are due to local causes. These free movements of the atmosphere, north and south, are not interrupted in the United States by mountain-ranges running east and west, but, on the contrary, they run in a northerly and southerly direction, nor are they for the most part so high as to prevent the surplusage of clouds loaded with moisture from passing over from one slope of a mountain-range to the other.

The Panoramic View.—To have a panoramic view of the territory of the Union, let the spectator imagine himself sufficiently high in the air directly above the point where, in the Gulf of Mexico, the Tropic of Cancer intersects the ninety-first or the fourteenth meridian. Directly north would be in view the Mississippi River, with its head-fountain near the northern boundary; he would also recognize on both its sides, east and west, numerous tributaries, hastening in a more or less southerly direction to mingle their waters with those of the great river. The impression·made upon his mind of the valley of the Mississippi would be near akin to that of a vast landscape-painting, tilted on the north side and facing the sun; framed on the east by the Alleghany Mountains, and on the west by the Rockies.

To the northeast he would see the shore of the

Atlantic, while almost parallel to it would be in view the range of the Alleghanies; the latter, commencing in foot-hills within one or two hundred miles of the north shore of the Gulf, and extending northeast for a thousand miles or more. Still northeast of their termination would be seen the Catskills, and yet farther north the Adirondacks, stretching from the Mohawk to the St. Lawrence; and east of the latter would come in view the range of the Green Mountains, while to their northeast would appear the White, not in a regular range, but solitary and alone, standing in separate masses, though proudly looming higher than their neighbors.

To the northwest, but much farther from the Pacific than are the Alleghanies from the Atlantic, come in view the Rocky Mountains—the backbone of the continent—commencing near the shore of the Arctic Ocean, and stretching southeasterly through the British possessions and across the United States into Mexico, on their way to Patagonia. On the parallel of 40° the Rockies are distant from the Alleghanies nearly fifteen hundred miles. West of the former—more than six hundred miles distant—appear the Sierra Nevada, in their northern portion called the Cascades; and still nearer the Pacific, and parallel with it, come in sight a series of highlands, known as the Coast Range. Between the latter two are extensive and remarkably fertile valleys, while the Great American Basin lies between the Rockies and the Sierra Nevada.

Utility of these Mountains.—The mountains of the United States are by no means waste; they exert an influence in the economy of Nature that increases the fruitfulness of the surrounding region, and also in various ways promotes the general healthfulness of the inhabitants. In the West they are nearly all storehouses of the precious and other metals, or covered with forests, except the highest points of the Rockies or the Sierra Nevada; while in the East the White, the Green, the Adirondacks, and the Catskills are covered for the most part with valuable forests, though not specially rich in minerals, except the southwest spur of the Catskills in their anthracite coal. The Alleghanies, however, are quite an exception; for, while covered on both sides—foot-hills and all—to their very summits, and from end to end with heavy forests, they also abound nearly to the same extent in coal and iron.

THE ancient nations that were the most advanced in civilization and progress, and who lived in the interior of the continents, occupied the most fertile districts, where they could obtain an abundance of food; witness the great empires that succeeded one another in the fertile plains through which flowed the Euphrates and the Tigris, and the empires that flourished in the valley of the Nile. Other peoples clustered around the shores of the great waters, and a few of the enterprising ones became leaders in using them as a highway.

Maritime Nations.—On the eastern borders of the Mediterranean Sea the Phœnicians were the first to venture out upon its waters, and, in time, planted colonies upon its southern shore, from whom sprang the Carthaginians. The latter prosecuted commerce in every direction, even out into the Atlantic and around to the British Isles. Meanwhile on the north shore of the great sea were the Greeks, having a country penetrated with numerous bays and inlets, thus making intercourse between themselves easy and also with the outside world; on the same shore were the Romans, and afterward the Venetians and the Genoese, who, having similar facilities, made the same sea their highway, and, in their interchanges of products between the nations, extended civilization and commerce. But the finest harbors of Europe are found amid the waters that sweep around the British Isles and along the north shores of France and Germany, and into the Baltic, and return by way of the south shores of Sweden and Norway. This portion of Europe has an immense advantage in the comparatively great length of her coast-line, thus extending around the bays and estuaries of these much-indented shores. England furnishes the most remarkable instance in modern times of great progress in commercial enterprise. No doubt her people were stimulated to trade in consequence of the facilities afforded them in having numerous and good harbors.

Comparison of Coast-Lines.—Let us compare this resource of the United States in proportion to their entire territory with the whole of Europe, and not alone with clusters of good harbors, bays, and inlets, in a comparatively small section. Prof. George Grove, an English authority on geography, estimates the coast-line of Europe at 19,500 statute miles, of which 3,000 along the Arctic Ocean is deemed unavailable for commerce. (*Primer on Geog.*, *p. 61.*) In order to make a correct comparison between the extent of the coast-line of the United

States and that of Europe, we must take *all* of the former, since it is *all* available for commerce, and of the latter *only* the portion that can be thus utilized—that is, 16,500 miles.

The coast-line of the United States—omitting Alaska—extends from the mouth of the St. Croix on the east border of Maine on the Atlantic, taking in all the indentations, estuaries, sounds and bays, and rivers to ports of entry, to the mouth of the Rio Grande on the Gulf of Mexico; and on the Pacific, from half a degree below the Bay of San Diego on the south to the Straits of Fuca on the north. Taking as a basis the available coast-line of Europe, 16,500 miles, and its area in square miles, 3,700,000, and the result will be one mile of coast-line to about 224 square miles of surface. Europe, in this respect, is the most favored of any division of the Old World; it having, when all the coast-line—19,500 miles—is reckoned, to each mile 190 square miles of surface; Asia has to one mile of coast-line 469 square miles of surface; Africa, one to 895; South America, one to 434; and North America—when all its coast-line is reckoned—one to 265. (*M. Reclus, "Earth," vol. i, p. 70.*)

United States Survey.—In a communication to the author from the United States Geodetic Survey-Office, Washington, and signed by Mr. C. P. Patterson, superintendent, is the following statement: " The greatest length of the shore-line (of the United States), measured in steps of but one mile, and in-

cluding all bays, *tidal rivers*, islands, etc., would probably exceed 150,000 miles, exclusive of Alaska." As
reported on this subject by the chairman of a com-
mittee of the Senate, the United States had more
than 17,000 miles of sea-coast to provide with signal-
lights, to which statement some of the senators de-
murred, holding the opinion that the number of
miles was greater.   To this must be added the
shore-line of the Great Lakes, at the least 1,000 miles.

The shores of the tidal rivers of Europe—that is,
those up which the tide flows—are reckoned as
coast-lines as far up as the tide influences the stream.
Thus Prof. Grove speaks of the great utility of
that "useful, civilizing river," the Thames, and its
flow of tide in promoting commerce.   The tidal
rivers of Europe are so small, in length, when
compared with the Mississippi, which is also tidal,
that it would seem fair to take the shores of the
latter and its tributaries up to their highest port of
entry.   Before the era of steamboats, ships cleared
from Pittsburg, at the head of the Ohio, for European ports.   From these interior ports of entry boats
pass down to tide-water to-day, and there transfer
their cargoes to ocean-going ships, and, in doing so,
the Mississippi affords facilities for commerce as
truly as the Thames, though more indirectly.   The
shore-line from these ports of entry on the Mississippi and its tributaries to the Gulf is estimated to
be at least 10,000 miles, which, with the ocean and
lake coast-lines, make 28,000, and this gives to the

United States, excluding Alaska, one mile of coast-line to about 108 square miles of surface—a result showing the numerous facilities that Nature has afforded the United States for promoting inter-course and commerce with other nations, as well as an almost unlimited means of communication among the States themselves. We may be able to appreciate these advantages of to-day, but can only imagine how much they will be enhanced in the future, as the ratio of national progress—moral and material—continues to increase from generation to generation, while over all pervades the stimulating and benign influence of a Christianized civilization.

COAL.

The Value of Coal.—It may be interesting to the American reader, on being introduced to the coal-fields of his own country, to glance at the untold value of that wonderful mineral, which the Creator has laid away in storehouses so immense. Of all the minerals taken together, it is the most important in its various applications, inasmuch as upon it depends man's ability to utilize so many of the others. In enumerating only partially what coal does for man's comfort and progress, we are amazed at the magnitude of the blessings it confers. Our great deposits of iron-ores, were it not for coal to smelt them, would be stripped of nearly all their value, for the labor would be increased immensely, in order even to produce the comparatively small amount of iron and steel that enters into the common implements of the farmer and the mechanic. What, then, shall be said of the immense quantities of iron needed for the great manufacturing industries of the land? It generates the steam-power that drives the machines, great and small, which in untold varieties produce articles that minister to the happiness of the people, while, with its assistance from the crude ore, molded by the genius of man, come the locomotive and the steel rails over which it glides so easily, and the steam-engine on board the iron ship, to transport products, and, in return, bring the necessaries so essential to the comfort of civilized communities. Shall we not also recognize the numberless blessings—each one small in itself, but in the aggregate so many—that it confers upon the people at their homes, in its domestic uses, in the genial warmth it gives, or, when needed, turns darkness into cheerful light?

The Carboniferous Age.—Geology treats of the various periods in which the earth underwent changes in its structure, but we shall notice only the one that has had a direct bearing on the formation of coal— the Carboniferous or Carbon-producing age. "Carbon is an elementary substance in Nature, which predominates in all organic compounds." To the latter belong plants of every description, from the tiniest spear of grass to the largest tree; these all have an organic structure, by means of which they derive from the earth, the water, and the air, the various substances—plant-food; by the latter's assimilation, they promote their own growth. These classes of vegetation or plants, one and all, are composed of carbon, combined with other substances, such as bitumen and the ingredients that produce gas, and which are termed volatile, as, under certain conditions, they can be driven off—when they are thus removed the carbon remains—the latter being the basis of charcoal, while also it enters largely into mineral coal. The Carboniferous age produced an enormous amount of vegetation, the remains of which, in the laboratory of Nature, were afterward converted into coal; the *rationale* of the process we leave to the chemist and the geologist to explain. "The Carboniferous period is probably not more than one thirtieth part of the world's history as recorded by geology. In the strata of that period are virtually all the coal-measures, or at least nine tenths of all the workable coal, in the world, as it was essentially the *coal-bearing period.*" (*Le Conte's Geol., p. 334.*) Coal-measures consist of strata of sandstones, shales, and slates, ledges of limestone and other rocks, also sometimes beds of iron-ore; amid these strata are interspersed seams of coal, but of various thicknesses.

The Peculiar Vegetation.—During this important period the sunshine and the original warmth of the earth stimulated an enormous growth of a vegetation so peculiar that it could not exist under other conditions. In connection with the coal derived from the remains of these immense forests, thickets, and undergrowth, are found about seven hundred species of plants, of which the greatest number belong to the fern family. They grew in depressions or basins that appear to have had no outlet, but whose bottoms were marshy or saturated with water, and some species were so tall and large—

even sixty feet high, and three or more feet in diameter at the base—that geologists designate them tree-ferns; but their texture was so soft that their trunks are usually found flattened out in the coal-measures. There were also large vines with enormously ex-

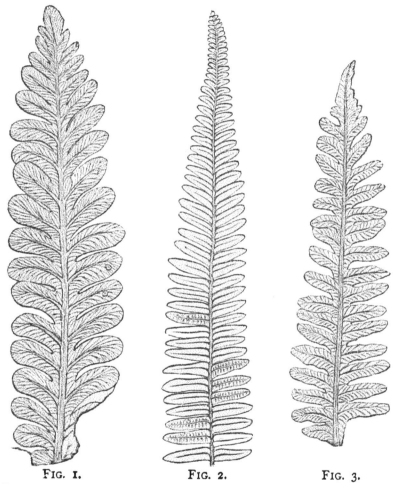

FIG. 1.   FIG. 2.   FIG. 3.

FIGS. 1, 2, 3.—COAL-FERNS : 1. Callipteris Sullivanti (after Lesquereux). 2. Pecopteris Strongii (after Lesquereux). 3. Alethopteris Massilonis (after Lesquereux).

panded roots and immense foliage; gigantic flags and rushes, and ferns of a smaller kind, that thickly studded the bottom. Here was a species of moss, named the *club*, from the blunt and rounded form of the extremities of its branches, that sometimes grew fifty feet high. None of these were "solid wood, but sappy, full of carbon or resinous and oily juices, containing more of the solid matter of coal than our most solid trees of to-day." (*Coal, Iron, and Oil, p. 68.*) These all, of course, grew partially under water; the evidence of this statement is derived from the remains of stumps and roots that are found in the impervious clay which constituted the bottom of the marsh. This forest, so singular in its characteristics, as its leaves matured, shed them into the water, there to be preserved till other influences transformed them into coal; first taking the form of peat, as at present found in bogs; then, perhaps, being subjected to a higher temperature and pressure from the outside, it became coal. Experiment has shown that purely vegetable peat, when submitted to hydraulic pressure, becomes virtually coal. To produce *twenty* feet of coal—the average depth of that in the United States—would require, it is estimated, a deposit of peat one hundred and twenty feet in thickness. How can we comprehend the amount of pressure that could reduce this mass to one sixth of its original size? Prof. Dana thinks that "hard wood would be reduced three fourths in weight and seven eighths in bulk to form ordinary bituminous coal."

Comparison of Forests and Temperature.—We

have but little conception of the magnitude of
that forest which once pervaded the districts of
the United States
wherein we now
find coal. A par-
tial illustration may
be derived from the
dense and tangled
forests and under-
growth now exist-
ing in the valley of
the Amazon, as it is
within a tropical
climate, and the
water is so tepid as
greatly to stimulate
a vigorous growth
during the entire
year. It may be
noted that hard
woods, now so use-
ful, are found on
the Amazon, while
the fiber of the
tree-ferns and the

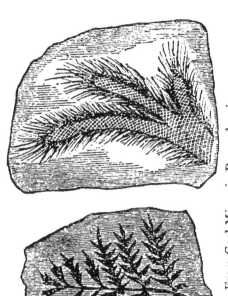
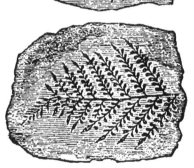

FIG. 4.—FOSSIL PLANTS: From Coal-Mines in Pennsylvania.

undergrowth of the unique forests of the Carboniferous age, were so little compact as to be worthless in
this respect; the latter's *mission* was to confer untold
benefits upon man when he should appear upon the
earth. Like that on the Amazon of to-day, this

strange growth knew no intermission, and we infer, from the data given, that the vegetation of that period was much more dense than that within the tropics of the present time.

Numbers of the species of ferns, whose remains are found within the coal-measures, are extinct in the regions where the coal was formed, but still exist within the tropics. They show, however, that their life and enormous growth depended upon the conditions of heat and moisture, and to an extent even far beyond what prevails now under the equator. In proof of this may be cited the fact that within the tropics are found representatives of the vegetation of the Carboniferous age—certain flags or reeds, that then grew to be *fourteen inches* in diameter and of proportionate height, but are now only *one half* inch in thickness; and tree-ferns, whose stems loomed up nearly sixty feet, but to-day their representatives are seen in the common ferns of our woods, while certain varieties of the club-moss, that grew to the height of . fifty feet or more, are now quite insignificant in size. The coal species of ferns are very similar throughout widely separated regions. That the sunshine was more effective in that age than now is improbable, since the vast amount of moisture that saturated the atmosphere would greatly absorb and lessen the warmth of its rays before they reached the earth. The comparatively high temperature of the tropics seems, in our day, to have little effect upon the cold of the higher latitudes, yet in the Carboniferous age was a vegetation of the same kind of ferns, etc., of sufficient growth to produce seams of coal in the Arctic Archipelago, on Melville's Island—within fifteen degrees of the pole—and also on the Arctic coast, at Cape Beaufort, in Alaska. Here must have been a temperature so high as to counteract the coldness incident to the sun's retiring during the winter months, but also to keep up the warmth the year through. Is not the solution of this phenomenon—the extraordinary putting forth of the energies of Nature in this wonderful growth—to be found in the theory that in that age the independent warmth of the earth, though in the process of cooling, kept the water so tepid that vegetation grew with very great rapidity? " The climate of the coal period was undoubtedly characterized by greater *warmth*, *humidity*, *uniformity*, and a more *highly carbonated* condition of the atmosphere than now obtains." (*Le Conte's Geology; Coal-Measures, etc.*)

The Reference to **Geology.**—We leave to the geologist to explain the composition of the coal-measures, and the transformations that have taken place in their midst; the numerous and varied strata of different thicknesses, such as that of sandstones and other kinds of rock, of shales and of slates, ledges of limestone, deposits of iron-ore, and seams of coal, though quite separate one above another, and sometimes of different kinds, as of bituminous and cannel; some seams sufficiently thick and some too thin to be made available.

The Great Basins.— The surface of the earth, before it was disturbed by internal forces, was com-

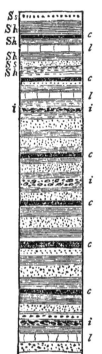

paratively smooth, yet it had rough places, but not nearly so great in proportion to its diameter as are those on the rind of an orange. Into these lower places drained the water from the surrounding higher ground. " The seed was in the earth," says a very ancient authority, and within these basins or depressions sprang up spontaneously a peculiar vegetation, already noticed. The indications are that in the Carboniferous age by far the greatest of these depressions within the territory of the United States extended in its original state, from the eastern edge

FIG. 5. — Ideal Section showing Alternation of Different Kinds of Strata. *Ss*, sand-stone ; *Sh*, shale ; *l*, limestone ; *i*, iron ; and *c*, coal.

and farther south of what we now term the anthracite field in Pennsylvania, westward to hundreds of miles beyond the line where the Mississippi now flows.

The Internal Movements.—It is evident that coal originally took the bituminous form, and from it, under certain conditions, came the anthracite. When this great marshy area had been covered, we know not to what depth, by a vegetable deposit, and was prepared, perhaps, in the form of peat to be transformed into coal, then through its eastern portion internal forces, by a lateral movement toward the northwest, shoved up the lower strata and folded them together with the beds of the peat or incipient coal. This convulsion was accompanied by heat, and in the folding of the strata so much pressure was brought to bear upon the mass that it was condensed into coal. Meantime the heat was not sufficiently great to burn the latter, but only to drive off or evaporate its volatile elements—the bitumen and the ingredients that constitute gas—and the result was anthracite. This internal movement extended in a lateral direction for nearly nine hundred miles, terminating on a line—the crest of the Alleghany Mountains—running northeast and southwest. Prof. Le Conte, in his *Geology*, says: " The sediments which had been so long accumulating in the Appalachian (Alleghany) region at last yielded to the slowly increasing horizontal pressure, and were mashed and folded and thickened up into the Appalachian chain and the rocks metamorphosed. In America this chain of mountains is the monument of the greatest revolution which has taken place in the earth's history." The folding of these strata one geologist illustrates by comparing the movement to that of a mass of dough being pressed from the sides, which would cause portions of it to fold one over another; they are also characterized by another geologist, Prof. Guyot, as "wrinkles on the face of Mother Earth." These basins or depressions, with their deposits of vegetable matter, collected during the Carboniferous age, have been illustrated by finding a counterpart in the famous Dismal Swamp of our own times, lying partly in Virginia and partly in North Carolina. In this are imbedded vegetable matter that may have been accumulating for thousands of years. It is known to be from twenty to thirty feet deep, and within it are found trunks of trees, which are often dragged out of their hiding-places and utilized as lumber. As far as we know, it is only needed to transform this immense deposit into coal, to furnish a sufficient amount of heat and pressure.

After noticing the coal-fields found east of the Alleghanies, we will treat of the effects of this internal movement on the west side of those mountains, and also of the coal deposits found in the valley of the Mississippi.

III.

IN treating of the coal-fields of the United States, we propose to notice first those on the Atlantic slope, which are much smaller and somewhat isolated from the more important fields of the Union. In New England coal is found near Worcester, Massachusetts, and in Rhode Island. These coals are in limited quantities, and very hard, partaking of the nature of anthracite. Vegetable remains are found in the coal-beds of Rhode Island, which indicate that the coal is the product of the Carboniferous age, the same as that of the great central fields. No such remains have been discovered in the coal found in Massachusetts; perhaps they have been destroyed by intense heat. The anthracite of Rhode Island appears to have been subjected to very high temperature as well as to very great pressure. The veins of this coal are quite irregular in form, and in thickness range from a few inches to twenty-three feet. The deposits are limited in size, while the quality of the coal is not so good as that found in the anthracite regions of Pennsylvania, the former having only 77 to 84 per cent of carbon, while the latter has 95. The coal thus being of less commercial value, the mines, as far as the general use is concerned, are virtually abandoned. "The beds are too unreliable and irregular to permit the production of coal with economy, or in competition with the mining operations of Pennsylvania."

Virginia and North Carolina Coals.—In the State of Virginia, thirteen miles west of the city of Richmond, is found a coal-field that in its extreme length north and south is about thirty miles, its greatest width being eight. It crosses the James River and extends south to the Appomattox. It is an interesting fact to the geologist that this bed of coal is in a fracture or trough of granite, upon which it rests without anything intervening except here and there a few inches of coal-shale. Only about one half of this coal, owing to the undulations of the granite, is available for working. There are, also, several small deposits of coal in the vicinity, in all estimated at 185 square miles. These coals are white ash, and highly bituminous, but they vary much in quality, some being quite impregnated with sulphur, while others are deemed good coal. In the Richmond coal-field is a singular seam of natural coke from five to six feet in thickness. It appears like coal from which has been driven off the volatile ingredients, while there was sufficient pressure to prevent the cellular texture of ordinary coke, but still retaining the carbon of the original coal.

It is interesting as a matter of history to notice that the Richmond coal-field was discovered and operated as early as the middle of the last century, and afterward there was an active trade in coal from Richmond to Philadelphia, and a more limited one with New York and Boston.

Piedmont **and Dan** River **C**oal-**F**ields.—This coal-field has an area of about twenty square miles; is located due west from that of Richmond, with which it lies parallel. The seams are generally thin, ranging from six to thirty inches, except in the vicinity of Farmville, on the Appomattox, where there is one seam sufficiently thick to be workable to some extent. But all these small deposits are irregular, and the coal dips in almost every direction, as this district appears to have been much disturbed by internal action.

In the vicinity of Danville, on the Dan River, and southwest from the Piedmont, is a coal-field known as the Dan River. It lies partly in Virginia and partly in North Carolina. It is evidently a continuation of the Piedmont field, the connection, however, having been broken by internal convulsions. The area is between twenty and thirty square miles, but upon the whole it is quite unimportant, as the seams of coal are generally so thin as to be unavailable for mining.

Virginia Anthracite. — Before leaving the Old Dominion we will briefly notice a coal-field within her borders, that has not yet been fully developed, but is full of promise. This field consists of several basins, located on the west side of the Shenandoah Valley, and at the east base of the Alleghanies, about fourteen miles west of the town of Harrisonburg. The coal found in these basins, which lie parallel to the mountains, may be deemed medium anthracite, as it contains less carbon than the anthracite of Pennsylvania, but more than the semi-bituminous. Here are several basins from which, in different localities, coal has been taken, and, on being tested, was found to contain 86 to 89·5 per cent of carbon. The presumption is that, when the mines are fully opened, the quality of the coal will be found much improved. This coal territory extends about twenty-six miles, and averages three in breadth, while there are at least three seams, ranging from four to eight feet in thickness, along a line of eight miles. The coal they produce compares favorably with that obtained in Lykens Valley, at the extreme southwest end of the Schuylkill field in Pennsylvania. It would appear that in a southwest direction from the main anthracite fields, in the latter State, the carbon in the coal diminishes from 95 per cent to 88 in Lykens Valley —perhaps the pressure upon it was not so great. The Virginia anthracite burns freely, and has a heating power proportionate to the amount of carbon it contains. The coal-seams in these basins have a dip toward the southeast, and have been sometimes folded by the great convulsion similar to the anthracite seams in Pennsylvania.

Deep River Coal-Field.—In a southeast direction from the Dan River field we find in North Carolina a coal-field having an area of about sixty square miles, where the seams are workable, and known as Deep River. It lies in a trough or long basin, which is depressed from one to two hundred feet below the general level of the country. Here are found bituminous and semi-bituminous coal, the latter shading off almost to anthracite. The bituminous is of excellent quality, making heat sufficient for the forge; it is clear, burns easily, free from sulphur, and is used for making gas for illuminating purposes.

New River Coal-Field.—This field is in Southwestern Virginia, on the river from which it takes its name. The latter, having cut its channel from the south end of the great Valley of Virginia through a break in the Alleghany Mountains, and by joining the Kanawha, its waters find their way to the Ohio. The seams of coal in that region are generally thin and crushed, and it is somewhat impure, though frequently similar to anthracite in character. This field or trough is not more than one thousand feet wide, but it extends a number of miles, and within it are several beds of available coal, which is quite free from sulphur, but contains an unusual amount of earthy matter. It is inconsistent with our design, in giving an outline of this branch of the Nation's resources, to cite in detail the various theories of geologists. They agree, however, in thinking that the deposits of coal from Richmond southward along the eastern slope of the Alleghanies are all of a more recent formation than that of the Carboniferous age, which produced the great fields of the Union.

IV.

THE ALLEGHANY ANTHRACITE COAL-FIELD.

THERE are four extensive coal-fields in the United States that are the production of the Carboniferous age. Of these, by far the most important is the Alleghany. This field extends in a northeast-south-west direction, and is in length about 875 miles, and of varied width, from thirty miles in the extreme southern end to 180 in its greatest width—from Cumberland, Maryland, to Newark, Ohio. The entire area of the field is 60,000 square miles, and within it are found the best classes of coal in the Union. Through its east middle portion run, in the same direction, the Alleghany Mountains. The latter extend from near the New York State line till they bluff out in Northern Georgia and Alabama. In the northeastern portion of this field are located the districts that contain anthracite coal. Though in the midst of mountains, the latter are the south-western projection of the Catskills. (*Coal Regions, etc., p. 11.*) Here are numerous deposits of anthracite coal, contained in troughs or basins, and on the west side of the Alleghanies comes the great

bituminous field, lying partly within eight States;
while intermediate between the former two are the
semi-bituminous deposits. The comparatively small
field of anthracite—*only 472 square miles by measurement*—excels, in point of mineral wealth, any area
of similar extent in the world. The anthracite coal-
mines of Pennsylvania are more valuable than the
gold and silver ones of California and Nevada. This
field was, evidently, once connected with the great
continental one to which allusion has been made
(p. 13), but was separated from it by the convulsion
that pushed up the Alleghanies, while apparently, in
the process of this internal movement, this portion
of bituminous coal was transformed into anthracite.
The coals of the Alleghany field are of three classes
—the bituminous, the original form, and from which
are derived the semi-bituminous and the anthracite.

The Parallel Movement.—It is remarkable that
nearly all the basins of coal east of the Alleghanies,
including those of Pennsylvania, Virginia, and North
Carolina, are parallel to one another and to the mountains themselves. The evidence is presumptive that
the great internal movement which pushed up the
mountains came from the southeast, in a wave-like
motion, and moved and crushed up the strata of the
crust of the earth toward the northwest; but the
resistance—the strata, perhaps, being deeper and
stronger—was so great on one line that the movement was arrested, and its force expended itself in
pushing up the mountains themselves, and only tilted

toward the northwest the strata which were adjacent on that side to the line mentioned.

Pennsylvania Anthracite.—We come now to treat of a coal-field—the anthracite—much limited in area when compared with the vast extent of the bituminous, but of remarkable value. A population of about 13,000,000, and occupying an area of 300,000 square miles, virtually receive their coal-fuel from this district, containing only 472 square miles. The volatile parts having been previously driven off in the transformation of the anthracite from the original bituminous, the former burns with scarcely any flame, and leaves a small amount of ashes. It is often designated by the color of the ash it makes, as *white* or *red*. The latter owes its color to the presence of the oxide or rust of iron. The white ash usually has 90 to 95 per cent of carbon, and the red ranges from 85 to 90. The latter burns more freely; both have great heating power, and leave but little ashes, while in domestic use both kinds are much valued for their cleanliness.

Mr. James Macfarlane gives the total area of the Pennsylvania anthracite coal-basins as follows:

1. Southern, or Schuylkill basin.............. Area, 146 square miles.
2. Middle, the basin Shamokin (50), Mahanoy
    (41), and Lehigh (37)................. " 123 " "
3. Northern, Wyoming and Lackawanna basin.. " 193 " "
    Total........................... " 472 " "

Discovery of Coal.—Before entering upon a description of this remarkable region, the reader may be interested in the stories of the two discoveries of coal, which arrested attention, and also the incidents attending its introduction to the public. Tradition tells that Nicho Allen, a hunter, encamped one night, in 1790, in a mountain in the Schuylkill region. He kindled a fire with wood amid some black stones; having gone to sleep, he was waked by the heat. We may judge his astonishment on finding the black stones all in a glow. He was much alarmed, thinking he had set the mountain on fire. This story was noised abroad, and the following year (1791) it happened that another hunter, Philip Ginter, was out on a mountain in the Lehigh district, and came upon a tree that had been blown down, the roots of which had upturned black stones. Having heard of the previous discovery, just mentioned, Ginter suspected these black stones were of the same character, and he carried specimens of them to one or two intelligent gentlemen, who rewarded him for showing them the place where the stones were found. This spot was on Mauch Chunk, or Bear Mountain, where the village of Summit Hill now stands. Here was a phenomenon seen nowhere else—a mass of pure coal, fifty-five feet high, and standing above-ground. It had withstood the heat and frost and rain-storms for thousands of years. Meanwhile much of the strata above, and all that was around it, had been eroded and carried away. This plateau of coal, in area between thirty and forty acres, stood alone on the top of the mountain.

(*Coal, Iron, and Oil, p. 119.*) Years afterward, here commenced the mining of anthracite coal by simply quarrying it out of this pure mass, which yielded 85,000 tons to the acre. It is worthy of note that this mine accidentally took fire in 1857, and is still burning (1887). The approaches to the place were, at the time of discovery, almost inaccessible to individuals, not to speak of the difficulties of transporting the coal.

Incidents.—It was a long time before the people learned to use and appreciate the excellent properties of anthracite, and, in consequence, it was years before there was a market for such coal. Judge Fell, of Wilkesbarre, in 1808, for the first time successfully burned anthracite in a grate, though at the time the experiment attracted little attention. Four years later Col. George Shoemaker took to Philadelphia, for sale, nine wagon-loads of anthracite, from near Pottsville, in the Schuylkill coal-fields. Bituminous coal, from Richmond, Virginia, had been used to some little extent in that city, but the good people were suspicious of this black, hard, and very peculiar stone—a stone-coal. Very few purchased it, and they in small quantities, but were unable to make it burn. The colonel was denounced as a swindler and a cheat by the indignant purchasers; and, indeed, at their instigation, warrants were issued for his arrest. However, he, learning of these legal proceedings, eluded the officers of the law by slipping away. Soon after the colonel's adventure a manufacturer of iron in Delaware County used anthracite as an experiment in his business, and astonished the world by proclaiming its excellent properties as a heater. Tradition tells that the workmen labored during the forenoon of the day to make the coal burn, by continually poking it, but in vain. In despair they closed the doors of the furnace, but left the draught open, and went to dinner and leisurely returned, but expecting that meanwhile the fire had gone out; but, to their surprise, they found the whole furnace in a glow! The *secret* was discovered. The news soon spread, and it was inferred that untold wealth lay hidden in the mountains of the State.

The Speculators.—Soon after this announcement the coal-trade began in earnest. Speculation was roused, and capitalists ran wild. Canals were projected to bring the coal to market, and railroads followed—then quite a novelty in the mode of transportation—for the same purpose. Millions were invested in the Schuylkill region alone. Laborers and mechanics, of all kinds and grades, and from all quarters and nations, flocked to the coal-fields, and for a time found ready and constant employment, at exorbitant wages. Towns speedily sprang into existence; many investors were disappointed, while some were successful. Every year more mines were opened, more iron-works were established, more improvements were planned, and more tons of coal sent to market. This feverish excitement gradually subsided, and coal-mining and transportation began to be conducted on business-like principles. The valleys of the Lehigh and the Schuylkill were the favorite fields of these operations.

The Schuylkill Coal-Field.—As already noted, Nature has divided the anthracite region of Pennsylvania into three districts, and, as the characteristics of the coal itself, the manner in which it exists amid other strata, the relative position of its different seams and of the surrounding country, are very similar, an outline of one district gives a general view of the other two. In noticing these coal-fields in their order, we will commence with that of the southern or Schuylkill. This, with the Lehigh basin, was the first to be developed, because, though after many disappointments, the owners finally secured, in 1825, a remunerative market in Philadelphia for the output of their mines.

This coal-field, as do the others, lies in a long, narrow trough or canoe-shaped basin; these several basins throughout the anthracite region are, for the most part, parallel to one another, though between them often intervenes a distance of a dozen or more miles. Mountains, parallel to one another, and running in a northeast-southwest direction, inclose the Schuylkill coal-field. The extreme length of this field is seventy-three miles—extending from the Lehigh to near the Susquehanna—with an average breadth of two, though in one place it is five, miles wide, for the basin is very irregular in its form and structure. "As a general description, it might be said that in the deepest part of this field there are fifteen coal-seams, each from *three* to twenty-five feet in thickness, in all one hundred and thirteen feet, of which eighty feet is considered marketable coal." (*Coal Regions, p. 21.*) These seams, if bared of other strata, would appear to the eye as running down the slope of one mountain-barrier to a great depth, from which point they would run up the slope of the mountain opposite and parallel. Or they might be compared to the ribs of a ship, starting from the keel, and extending up either side, and growing shorter as they approach either end. Thus the coal within these troughs has more seams in the middle, which become fewer and thinner toward the ends, while the bottom gradually turns up, "till the coal is pinched out." On the southeast side of the Schuylkill field the dip or inclination of the seams of coal is about 80°, while on the northwest side it is between 50° and 60°. They run down to a depth as yet unknown, but that point is estimated, perhaps from this dip, to be more than two thousand feet. Future generations will no doubt reach this lower seam in the pursuit of coal, as England has already gone down still farther, and, under the disadvantage of working in bituminous coal, the walls of which are less strong than those of the anthracite, while the seams of the English coal run from only three to four and a half feet in thickness. Although there are no horizontal seams, all those of the anthracite are not so regular in their dip as the ones composing the Schuylkill basin ; for illustration, in a portion of the Lehigh district, on Nesquehoning Creek, there have been so violent disturbances, by the internal wave-like convulsions, that the seams have been

FIG. 6.—Nesquehoning Basins (after Daddow).

folded over and against one another to a degree that they are almost perpendicular.

In these internal convulsions of the earth, mountain-ranges have been broken across and cut down by the erosions of time, to the valleys under which lies the coal, and through these cuts or gaps run streams ; the same depressions have been utilized for constructing railways for the purpose of bringing the coal to the outside world. We can not go into details, only observing that the whole anthracite region is so broken up, while the number of these passes and cuts is so great, that they give the landscape the appearance of a series of isolated mountains.

The Mammoth Seam.—In the Schuylkill basin at Pottsville—the center of the coal production of that field—is the deepest portion of the coal-seams. " The real wonder of this famous Pottsville region is the great ' Mammoth ' seam of coal, which is often as much as thirty, forty, and in some places even fifty feet in thickness. . . . This is the most regular and reliable of all our coal-seams, the most economical to mine and operate, and, from its size, the most productive." Throughout the anthracite region, owing to upheavals, the seams are much tilted, and, in consequence, the mining operations are more difficult and expensive than in the bituminous fields; but that is compensated by the value of the coal itself, and by its being in great demand. This immense seam, in the midst of others, pervades the entire anthracite region; and, when we come to treat of other localities, we shall find traces of it even there. The very large seams are the only ones worked at present, and, it is said, in a wasteful manner, the operators tacitly proposing to let posterity take care of itself; but the time will come, perhaps, in a few centuries hence, when the seams that are now less available will be worked. Improvements in machinery for getting out coal may be invented in the future, by which these smaller seams can be utilized, as those of similar size are to-day in England and on the Continent. That the reader may find on a map the localities of these remarkable coal-fields, we give the names of the principal mining towns: Thus, in the southern field, among many smaller ones, are Pottsville and Minersville. (*Appletons' Higher Geog.*)

The Middle Coal-Field.—This coal-field, though the smallest, is composed of quite a number of basins more or less detached from one another, such as the Shamokin and the Mahanoy, which are virtually one, though partially separated by a ridge. Their names are Indian, and derived from two creeks, tributaries of the Susquehanna. The western portion of the Middle field, the Shamokin, is twenty miles in length, with an average breadth of two and a half, the area being fifty square miles; the eastern portion, the Mahanoy, is twenty-five miles in length, with a mean width of less than two, and its area is forty-one square miles. In connection with these is also reckoned the Lehigh district, which numbers seven narrow basins, the combined area of which is thirty-seven square miles. The latter river is a tributary of the Delaware.

The Lehigh basins, though comparatively small in area, are very productive, and furnish an excellent quality of coal. It may be noted that the coals of all these mines have the same general characteristics; to some, however, slight advantages are conceded in their hardness and dryness, and a greater amount and purity of carbon. The coal from the Lehigh basins is in some respects very popular. It is specially valuable when an intense heat is required, as in the manufacture of articles made from pig or cast iron. For these reasons the use of that coal has extended more over the country than any other. The Mammoth vein in these seven basins is about thirty feet thick, and pure without a break, and noted for its dryness. The leading mining towns of this Middle field are Mauch Chunk, Tamaqua, Mahanoy City, Ashland, Shamokin, and Hazelton.

The Wyoming and Lackawanna Coal-Fields.—

This unbroken and largest and finest basin of anthracite underlies the beautiful Wyoming Valley, so famous in our history, because of a sad tragedy. (*Hist. American People, pp. 498, 502.*) Through the southwestern portion of this valley flows the main branch of the Susquehanna, which is joined within it by the Lackawanna, which flows through the northeastern part. Beneath the beds of these streams, at a distance of hundreds of feet, are the coal-seams. They extend, following the moderately crescent shape of the valley, for fifty miles without a break, with an average width of nearly four, the area of the coal-field being 198 square miles. The whole valley is completely shut in by mountains, through which, on the northwestern side, the Susquehanna cuts its way, making a channel down to the level of the surface of the valley, while the head-streams of the Lackawanna come in almost at the same level on the northeast. The coal-seams, apparently adapting themselves to the undulations of the valley, rise only to a moderate distance up the slopes of the inclosing mountains, but extend below to an unknown depth; in the central portion of the basin, that point is estimated to be several hundred feet beneath the level of the ocean. Excavations in pursuit of coal have not yet reached this lower region. The seams of coal in this northern field are quite regularly disposed and uniform, nearly approaching the horizontal position of the seams in the bituminous fields, thus facilitating mining operations. We may suppose that the wave-like movement in this northern field spent its force in throwing up the mountains, and was unable to fold the coal-seams over one against another, as in some of the other basins, or perhaps in the convulsion the coal strata sank down and remained unbroken, but merely bent upward at the sides and ends. The Mammoth vein in this field ranges from fourteen feet to twenty-four. The chief mining towns are Wilkes-barre, Pittston, Scranton, and Carbondale.

The Outlook—the Enterprise.—We have briefly noticed the anthracite fields of Pennsylvania, to whom nature has virtually given the monopoly of that class of coal. The usual estimated amount to the acre that is mined from the several workable seams is about 60,000 tons, though it is conceded that there is a vast deal of wastage in the mining operations as now conducted, which would amount to perhaps 25,000 tons more if all the coal were taken out. The time will come when the "Mammoth," and others of the larger seams will be exhausted, and from necessity those that are now deemed unprofitable to mine, because of their thinness, will be brought into requisitiou.

These anthracite fields afford examples of energy and enterprise that are not surpassed elsewhere in the Union. Railroads, after overcoming difficulties almost insurmountable, made their way along rivers and ravines, through gaps, and tunneling when they could not rise over summits, until they penetrated the inmost recesses of these mountains, once thought to be absolutely inaccessible to railways. The management of these roads gives employment to thousands, while almost everywhere in sight are steam-engines hoisting the coal to the surface or freeing the mines from water, or driving the immense machines that crush the anthracite into various sizes to prepare it for use, while thousands upon thousands of miners in the depths below, with picks and shovels, are digging the coal out of its original bed. Not long since (1885) there were 210 collieries at work in the southern and middle fields, and there were about 90 in the Wyoming and Lackawanna district, making in all about 300. From these statements we may form an estimate of the immense amount of labor required to furnish the people that are dependent upon these mines for their coal-fuel.

The Great Eastern Valley.—One of the most important natural resources of the United States is their abundance of soil that is fertile and available for cultivation. In connection with the coal-fields just described, it is proper to notice a depression or valley from ten to twenty miles wide, and about 1,500 miles long, that runs throughout, virtually parallel to the Alleghany Mountains and coal-fields, and to the Atlantic coast. It commences in New York State, northwest of the Catskills, crosses Pennsylvania, Maryland, Virginia, East Tennessee, and terminates in the State of Alabama. Into this valley enter many smaller ones, that are equally fertile. It is known by different names, such as the Cumberland

in Pennsylvania and Maryland, the Shenandoah in
Virginia, and the East Tennessee; the latter drained
by the Holston, the French Broad, and the Tennessee. It has one valuable characteristic throughout:
it is underlaid with limestone, whose elements im-
pregnate the soil, and make it pre-eminently the most
productive region in cereals east of the Alleghanies.
"All Southeastern Pennsylvania, owing to the pres-
enee of limestone and other softer rocks, possesses a
fertile soil, and has been justly called the garden of
the Atlantic slope." It is a striking feature of this
valley, when taken in connection with the proximity
of an abundance of coal, that in certain portions of
it, especially in Pennsylvania, are found large deposits of iron-ore, and often beneath the soil of the
wheat-fields may be found beds of this ore and ledges
of limestone. The proper combination of these three
materials, as a means to an end, is essential in making iron.

V.

SEMI-BITUMINOUS COAL-FIELDS.

BEFORE passing to the west side of the Alleghanies, where are found our most important bituminous coal deposits, we will briefly notice a series of basins which contain a coal that is unique in its nature, and characterized as semi-bituminous—a medium between the hard and soft coals. The semi-bituminous produces great heat, as it contains from 70 to 84 per cent of carbon, and has likewise sufficient bitumen or volatile matter to make a blaze. The latter trait renders it very useful in generating steam, as the flame passes through the tubes within the boiler; hence it is sometimes designated as *steam-coal*. The strata containing this coal were tossed up by the great convulsion, and heated and pressed, but not nearly so much as was the anthracite. The northern portion of this series of coal-fields lies west of the anthracite region, on the head-waters of the West Branch of the Susquehanna, the Alleghany, and the Juniata Rivers. These coal-basins lie high up in the mountains, some even 1,000 feet above the intervening valleys; the latter, in the course of many centuries, were made by the continuous erosion of the waters of the streams that now flow through them, and which scooped out and carried away the earth as well as the coal.

The Basins from Blossburg to Broad Top.— These deposits of coal lie in a southwest direction along the eastern margin of the Alleghanies, and only in this long, narrow region is this species of coal found. These basins are numerous, but often small, and, at intervals, extend from Blossburg to Broad Top, a distance of one hundred miles or more. Along this line are eight localities in which this coal is found, and in mining it are engaged nearly sixty corporations. Blossburg, on the north, has a coal area of 50 square miles, and Broad Top Mountain, on the south, has 80, while the remaining basins have altogether about 370 square miles. The fossils found in this coal show its origin to be similar to that in the neighboring localities.

The Blossburg mines are on the head-streams of Tioga River, which runs north to the Chemung, a branch of the Susquehanna. In these mines the seams of coal lie one above another, and many feet apart, and vary from three, four and a half, to five and a half feet in thickness, in all about thirteen feet.

Broad Top Mountain is quite singular in its general conformation, being a ridge encircled by valleys 1,000 feet deep. The ridge is eighteen miles long, by an average width of about four and

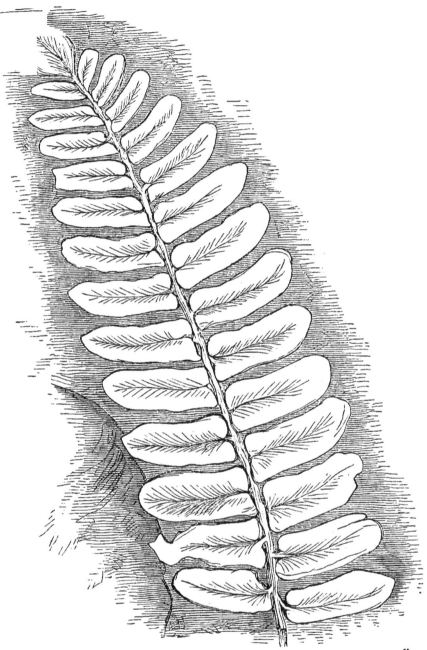

FIG. 7.—NEUROPTERIS FLEXUOSA: Literally, "Bending Nerve-Fern," so named from the Veins of its Leaflets. The Fossil Fern, here represented of the Natural Size, was taken from the Roof-Slates of Coal-Seam D, at the Mines of the Blossburg Coal Company, at Arnot, Pennsylvania.

a half. The coal-measures of these eighty square miles are disposed in six basins that are parallel to one another, amid which rises Terrace Mountain, a peak in height equaling the Alleghanies themselves. Near the summit of this peak remains a small round patch of the Pittsburg or Westmoreland gas-coal bed, a few acres in extent—the sole relic of that vast deposit " which once covered all that region."

This semi-bituminous coal, though having excellent qualities, such as being free from sulphur and of great heating power, is inferior in appearance, liable to crumble when exposed to the air, and become very fine, almost dust. It is easily broken, falling apart in cube-shaped pieces, and is decidedly dirty, soiling almost everything it touches. It is admirable for the use of the blacksmith, and for that purpose is sent to Canada, and is also used in the workshops of the railways.

Johnstown.—In treating of this interesting subjcet, we should not fail to note the many instances in which the Creator has placed near one another the materials that, when combined by man's ingeunity, confer untold benefits upon the race. A striking illustration of this is seen in that remarkable region around Johnstown, Cambria County, Pennsylvania. There, in the same mountain-side, lying in *five* horizontal seams, one above another, with spaces between of different depths, is found semi-bituminous coal. It is in seams ranging in thickness from three and a half to seven feet. In the same mountain-side, and interspersed with these seams of coal, are also found deposits of iron-ore and ledges of limestone. The iron-ore is in two beds, the main one overlying the highest workable coal-seam. The coal melts the ore; the limestone fluxes it—that is, when the former is intensely heated, the latter causes the mass to flow as a fluid. The only transportation required for these materials is the comparatively short distance from the mouths of these mines to the furnaces in blast.

The Cumberland Coal-Field.—The Cumberland district of semi-bituminous coal belongs properly to the great Alleghany coal-field, though it is separated from the latter by a mountain. There are composing the Cumberland coal district three prominent basins, all lying in a southwest direction, and separated from one another by the projection upward of mountain-ridges. The northeast basin, known as the Frostburg, is in length thirty miles, and has an average width of five. The middle basin has an area of 130 square miles, while the southwest one has 250; in all 530 square miles. The Youghiogheny River cuts its channel through the latter. Portions of the Frostburg basin lie within three States—the northeast in Pennsylvania, the middle in Maryland, and the southwest in West Virginia. The Cumberland coal is unusually free from sulphur, and contains from 13 to nearly 20 per cent of volatile matter, and from 72 to 83 per cent of carbon. The

main seam of the Cumberland field is nearly horizontal, and is fourteen feet in thickness—a relic of
the Pittsburg seam of the Alleghany bituminous
coal-field—while in the southwest basin, across the
Potomac, at Piedmont, in West Virginia, the same
seam, now sixteen feet thick, crops out 1,000 feet
above where runs the Baltimore and Ohio Railway.
The coal is taken out in large lumps. In color it
is of a jet-black, having a glossy appearance, but
withal becomes quite friable when exposed to the
atmosphere, and is soon reduced in size very much
in the handling necessary for its transportation.
Notwithstanding these drawbacks it is a remarkably
fine coal for generating steam, and therefore is very
much used on board of ocean-going steamers and
locomotives. It may be remarked that, owing to
the crumbling nature of the coal, and consequently
the lack of a strong, self-sustaining roof, the seams
can be very seldom mined to their full extent, as
in the anthracite mines, or even in the bituminous
fields farther west. Hence there is an unusual
wastage, and the fear is entertained that, if the
present system of mining continues, a half-century
will see these mines exhausted. It may interest the
reader to know that geologists state that "the
Frostburg or Cumberland coal-basin is the most interesting and important feature in the geology and
physical geography of the country." The entire
area of our semi-bituminous coal is 1,030 square
miles.

VI.

THE ALLEGHANY BITUMINOUS COAL-FIELD.

As intimated (p. 19), we come now to treat of the effects produced in the strata of the basin northwest and west of the Alleghanies by that internal convulsion, whose wave-like motion coming from the southeast crushed and folded up the crust of the earth on their southeast side, but whose movement was arrested on the line where now stand the mountains themselves. Along the entire length of that line— nearly 1,000 miles—for some unexplained reason, the resistance was too great to be overcome, and the result produced was that the whole southeast portion of the basin mentioned above was only raised up or tilted on the mountain's northwest side. This tilting appears to have had its hinge, so to speak, along the line of the Ohio River, that being the lowest part of the inclined plane, as under the latter's present channel there is evidence of a depression—the bed, perhaps, of a former stream. Geology tells us that "the Ohio, throughout its entire course, runs in a valley, which has been cut nowhere less than 150 feet below the present bed of the river. This discovery is due to the investigations of Prof. J. S. Newberry, now of Columbia College, New York city. In addition the influence of the internal move- ment of the crust of the earth seems for the most part to have stopped at that line, as under the prairies northwest and west of the river the strata appear to have been scarcely disturbed. Southeast of the mountains the strata were tossed and folded in comparatively small spaces, while on the west they were merely upheaved in vast areas, but not broken and crumpled. There was heat sufficient to modify the vegetable material in the strata, but not enough to drive off the volatile parts; and this mod- crate heat, combined with superincumbent pressure transformed—may we not say?—the peat into a bitu- minous coal, which in its general excellent qualities, and in consideration of its vast area, is the most remarkable field in the world. Another evidence of this line of depression is derived from the fact that all the original streams from the crest of the mount- ains, the highest part of the inclined plane, uniformly found their way to the Ohio. Thus from the high- est line or edge of this water-shed, which extends from Southwestern New York along the mountains to Northern Alabama, and thence west on a dividing ridge almost to the Mississippi, flow the Alleghany, the Conemaugh, the Youghiogheny, the Monongahela, the Kanawha, the Kentucky, the Cumberland, and the Tennessee, all into the Ohio, and all have their head- streams at or near the highest line of this inclined plane.

Rivers and Valleys.—The process of erosion or denudation commenced when from this crest the waters, at first in streamlets, then increasing in size by being joined by others, began seeking in a north or northwest direction the lowest part of this tilted plain. The main channels increased in size while their tributaries increased in number, and all cutting their way across the plain and down through the various strata of the coal-measures, scooping out by attrition the earth and the coal, and carrying both away. As a compensation to man were left instead fertile valleys, great and small, but all enriched by these streams and beautiful hills, in which are preserved the original strata, and out of whose sides often crop the seams of coal that have escaped the general destruction. It took ages and ages to excavate these valleys, and carry away their original contents; while in accordance with the nature of the strata to be overcome, some of the valleys are narrow, inclosed by steep and parallel hill-sides, and others are wider, and often have long and wide stretches of fertile bottom-land, bordering on these streams or rivers. All this has been done for the benefit of man; to be

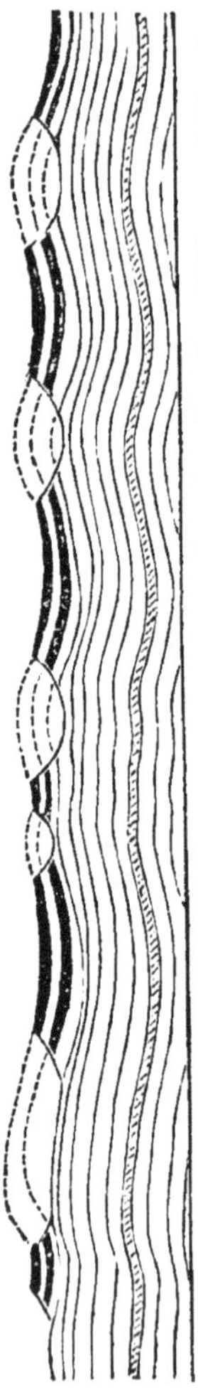

FIG. 8.—Section of Appalachian Coal-Field, Pennsylvania, showing Effects of Erosion on gently Undulating Strata (after Lesley).

sure, an immense amount of coal has been carried away, but no doubt there still remains sufficient for the wants of the Nation. In lieu of the apparent loss thus sustained, the process of cutting down the valleys has wonderfully increased the facilities for mining the coal, which in that region can now be reached on a level, rather than by shafts and expensive machinery, and extra labor, while the rivers thus made afford an easy means of conveying the coal where it is needed.

The Great Basin drained.—The great basin (page 18) was originally hemmed in on the southwestern side by a barrier that extended even west of where the Mississippi now flows. The weakest portion of that barrier was from the bluff, on which now stands the city of Vicksburg, to a similar one in Louisiana, distant sixty-five miles. The Vicksburg or eastern bluff is now two hundred feet high, and the western is a little lower. Prof. Guyot says, " The strata in the opposite bluffs correspond throughout in such manner as to prove their former continuity." Near the Vicksburg bluff the accumulated waters burst through the barrier, and then and there commenced not only the formation of the southern part of the channel of the Mississippi, but also its delta. The reader may ask, " Where did these waters carry the mass of the earth and strata that once filled the space occupied by the valleys of to-day ? " The same authority answers, " It is estimated that the sediment carried annually at the present time by the Mississippi to the Gulf is sufficient to cover one square mile to the depth of 268 feet." If that estimate is correct, after the banks of the streams and the hill-sides have been protected by verdure and by forests, how much greater must have been the erosion when the waters burst through the barrier! The area of the delta of the Mississippi is 12,300 square miles; that designates only the surface measurement, but we are unable to ascertain the depth of the ocean that had to be filled before the height of the present surface was reached, and therefore its cubical meas-urement can be only a subject of conjecture.

The Pittsburg Coal-Field and Seam.—The work-able seams of coal in this field are estimated in the aggregate of their depth to be from twenty to thirty feet; the lower ones being generally the thickest and the purest. We noticed in the anthracite region a seam of coal designated the " Mammoth," because of its immense size; so in the great bituminous field, west of the Alleghanies, we find another that holds a similar pre-eminence. The latter is known as the " Pittsburg seam," thus named from that city, so remarkably progressive in all the mechanical indus-tries that require a great amount of heat, such as the manufacture of glass, and of iron and steel. Pitts-burg has ever drawn its coal-fuel from that seam, as it underlies the hills in the vicinity, and, except to-ward the north, extends around for hundreds of miles. The coal crops out in different seams all along the hill-slopes that border the rivers—the Monongahela and its main tributary, the Youghiogheny, and also in the hill-sides of the numerous valleys scooped out by their several tributaries. The area of this coal-seam is about 18,000 square miles ; in connection with Northwestern Pennsylvania, it includes a portion of Northeastern Ohio, and the rest of Western Pennsylvania to the southern border, where it takes in West Virginia, and a portion of Western Maryland and of Eastern Kentucky. This seam at Pittsburg and its immediate vicinity is eight feet thick, but that depth gradually increases toward the southeast, till at the base of Chestnut Ridge, the most western subordinate range of the Alleghanies, it is from nine to eleven, and still farther in the same direction in the outlying Cumberland deposit, it reaches fourteen and sixteen feet in thickness. In its broad extent it manifests a wonderful uniformity of all the physical conditions under which, in the Carboniferous age, its vegetable ingredients were accumulated over its entire area. " This rich, solid, bituminous coal is very free from sulphur, and yields forty to forty-five cubic feet of gas to *ten* pounds of coal." About fifteen miles east of Pittsburg, along the Pennsylvania Central Railway, are several mines of this coal, whence it is transported to the Eastern cities for gas-making purposes. The station at Irvin, Westmoreland County, is the center of the most important gas-coal mining region in the United States.

It may be remarked that on an average of seventy-five feet above this main seam there is a deposit of limestone, the most extensive and valuable in that portion of the State, especially in the valley of the Monongahela and those of its tributaries. This may account for the remarkable fertility of the soil of that region, as it is impregnated with lime similar to that of the famous Cumberland Valley in Eastern Pennsylvania. "The Monongahela Valley has beautiful hill-slopes, backed by a great rounded, smoothed, cultivated up-country, of grain-farms and pasture-lands, the whole unbroken by deep ravines, and scarcely-indented valleys."

Horizontal Seams—Monongahela Valley.—It is noticeable that the various seams of coal in this region deviate but little from the horizontal; in consequence, the coal is easily taken out, and the mines are self-draining. The dip is so gradual that, starting on the southern line of the State, on the upper Monongahela, the seams, in about fifty miles, attain an elevation of at least three hundred feet above the river, where the coal crops out on the top of the hill. Now comes a break on a line running east and west, and the seam drops down to the water's edge. This is near the city of Brownsville. This striking feature arrested the attention of the celebrated English geologist, Sir Charles Lyell, when visiting, as a scientist, this locality. After expressing his astonishment at the magnitude of this seam of coal, he says, " Horizontal galleries may be driven everywhere at very little expense, and so worked as to drain themselves, while the cars laden with coal, and attached to each other, glide down on a railway so as to deliver their burden into barges moored to the river's bank." (*Elements of Geology, p. 392.*) No doubt Sir Charles's amazement was increased as he compared the ease with which this coal was mined with the difficulties of mining coal in his native land, where the seams are scarcely more than four feet in thickness, and the dip is so great that already a depth of 1,000 to 2,600 feet has been reached, thereby the labor of digging and hoisting the coal has been increased enormously, while, to free the mines of water by pumping, has equally increased the expense.

Coal beneath the Pittsburg Seam.—The gradual rising of the main coal-seam is repeated; for, starting from the river's edge at Brownsville, where it dropped, it gradually rises, and, forty miles distant, crops out 300 feet above the river, on the hills just south of Pittsburg. At that point it drops again, but much farther, so that the seams of coal are now from 140 to 180 feet below the surface, they underlying the city itself. These seams have been reached by means of shafts.

From this and other indications, there is reason to believe that these lower beds or seams of coal pervade that entire region, and that they are only portions of a series of coal-beds that lie far below the Pittsburg seam. Geologists say that in the vicinity of Wheeling, West Virginia, "the lower series of coal-beds are resting upon the conglomerate 600 or

700 feet below the Pittsburg seam." " The coal-beds of the lower coal-measures must underlie the whole of this vast area, at a greater or less depth beneath the general dip of the strata, westerly." (*Coal Regions, etc., p. 283.*) Similar seams of coal have also been reached in boring for salt-water near Greensburg and at Conemaugh, both east of Pittsburg, while on the upper Monongahela, some ninety miles south, near Clarksburg, a seam of coal, eleven feet thick, was passed through under similar circumstances. The latter was far below the river, while above it, in the hill-side, are several seams, one above another. Recently (1886), in drilling for gas in the vicinity of Fayette City, thirty miles south of Pittsburg, the drill passed through a seam of coal nine feet in thickness, and at a depth of 600 feet below the river.

The Characteristic Change.—About thirty miles southeast of Pittsburg, up the Youghiogheny, the quality of the coal begins to change, and in the succeeding twenty miles the seam itself grows larger, until it reaches from nine to eleven feet in thickness, terminating in what is known as the Connellsville coke basin. The latter takes its name from the village near its center, where the characteristics of this coal were discovered and first utilized in converting it into coke. This trough or basin extends along the western base of the Chestnut Ridge for almost ninety miles, from near the village of Blairsville to beyond Uniontown, in Fayette County, and to the West Virginia line. The area of this entire basin is estimated at 200 square miles. Strictly speaking, the Connellsville basin is not more than twenty or twenty-five miles in length, but its northern portion extends from Latrobe, on the Pennsylvania Central, "through a part of Westmoreland, and through Indiana and Clearfield Counties." The coal of this latter portion is deemed inferior to the Connellsville as a coking material. The Youghiogheny River cuts through the Chestnut Ridge near the latter village, and also, at that point, runs across the basin; this outlying ridge or range is lower than, but runs parallel with, the Alleghanies.

The Two Questions.—An interesting question arises in respect to this basin of coal being so much deeper and so much richer in the elements, *carbon and bitumen*, than the other portions of this great seam. Theory explains that in that part of the Alleghany field there was, in the Carboniferous age, a much denser growth than elsewhere of those specially juicy plants—ferns, for instance—and other vegetable materials, the outcome of which is that peculiar grade of coal. Another question may arise: What is the advantage of coking this coal before using it in furnaces or in foundries, where great heat is required? The ingredient in bituminous coal that produces nearly all the heat is carbon, while the bitumen and gaseous matter passes off in smoke and flame. By removing these volatile ingredients in the process of coking, we *obtain the carbon* in the form of coke, in which is concentrated the heating power of the original coal. This coke, when under blast, uniting with the oxygen of the air which is forced in, generates a heat so intense as to drive off or consume the extraneous matter in the ore, and likewise melt the iron itself most thoroughly. Analyses show that coke derived from raw coal having, say, 60 per cent carbon, will contain, in this concentrated form, nearly 90 per cent of the same.

The Excellence of the Coke.—This Connellsville coal has in it a greater amount of carbon than any other portion of the Pittsburg seam. It is remarkably free from impurities, such as sulphur and phosphorus, and the ingredients that leave ashes. In consequence of these traits, the coke obtained from it has superior qualities for smelting and producing a pure article of iron, and for preparing the latter to be easily converted into steel by the Bessemer process (p. 150). This coal averages about 60 per cent of fixed carbon, and this, combined with its purity, seems, when taken as a whole, the finest coke known. The latter " is compact, silvery, and lustrous "; is harder, heavier, and more condensed than ordinary cokes. " Its greater density enables it to sustain more weight in the furnace than any other quality, and it is therefore specially adapted to use in blast-furnaces and in the cupola of the foundry." These qualities cause this coke to be used more or less—it being also often mixed with raw coal—throughout the Western States, and in a moderate extent in the Eastern and Southern, in all industries in which great heat is required. As the materials for making coke are coextensive with the greater portion of our bituminous coal-fields of the Carboniferous age, we will mention the fact only when occasion requires us to state that the coal is not susceptible of being coked.

West Virginia Coal-Field.—This is one of the most interesting States in the Union in respect to its mineral wealth, as, in proportion to its size, it is the richest in coal, iron, salt, and oil. It has of coal, both bituminous and cannel, at least 16,000 square miles. The seams vary from three to twelve feet in thickness; in some instances, where they are one above another, the aggregate depth is as high as twenty-five feet; and, in mining the coal, all are very easy of access. Nine tenths of the valley of the Guyandotte River, in the southern portion of the State, are underlaid with seams of coal, while in the northern part, in the vicinity of Wheeling, in addition to the Pittsburg seam, another one of fine bituminous coal has been discovered 300 feet below the level of the Ohio. For aught we know, this seam may extend under that whole region; the indications are that it may be connected with the seam under the city of Pittsburg, already mentioned.

Coal is also found in great abundance in the hills along the upper Monongahela, in West Virginia, and its tributaries, especially in the vicinity of Clarksburg. Here the main seam is from ten to twelve feet in thickness, below which is still another, somewhat thinner, but of a coal more highly bituminous. Cannel coal also abounds in this region. In the central portions of the State, as in the valley of the Kanawha, along whose banks crop out large seams of coal, and on the hill-sides, especially on Coal River, one of the latter's tributaries. In Ritchie County, near the center of the State, is a remarkable mineral deposit, "a vein of asphaltum (compact native bitumen), *four and a half* feet thick, more than three fifths of a mile long, and of unknown depth. It fills a great fissure, which breaks through the rocks nearly perpendicular, and outcrops on the surface." Its analysis shows 55 per cent volatile matter, 42 per cent fixed carbon, and 3 per cent ashes. This is used for gas-making purposes. It is similar to the *Albertite* of New Brunswick (p. 99). The deposit itself has, no doubt, some connection with petroleum. The famous Peytona cannel-coal seam, about six feet in thickness, is on Coal River, twenty-five miles from its mouth. This coal is of an unusually excellent quality.

Eastern Kentucky Coal-Field.—This is a continuation, from West Virginia, of the Alleghany field. The northern limit of this coal-field is on the Ohio, opposite the town of Portsmouth, but it extends southeasterly along the western slope of the Alleghanies, here called Cumberland, and underlies the head-streams of the Big Sandy and the Cumberland Rivers, while in some instances the coal-seams are found half-way up the sides of the mountain. The entire coal area in this part of the State is 9,000 square miles. It has several seams, the main one being nearly seven feet of pure coal. Eastern Kentucky is not only rich in this coal-field, but in this connection it possesses also a vast amount of iron-ore, and an abundance of limestone.

The Tennessee Coal-Field.—The coal-seams of East Tennessee are peculiarly situated, as they occupy a mountain-plateau, 2,000 feet above the ocean, and from 900 to 1,200 feet above the valley of East Tennessee, already noticed as the southern end of the great valley on the Atlantic slope (p. 40). There are some remarkable geological features connected with this plateau, but to notice which is beyond the scope of this volume. (*Coal Regions, etc., p. 365.*) The eastern line of this coal-field of 5,100 square miles is parallel with Cumberland Mountain, and also of the valley, and, while that line is quite regular, the western side, though trending in a southwestern direction, is very irregular. The valley of East Tennessee is drained by the rivers Holston, French Broad, and Tennessee. It is a fine agricultural region, having a fertile soil, for here abound limestone ledges, as well as iron-ore and coal near at hand. On the southern border of the State, near the Georgia line, are the Chattanooga coal-mines, having the same peculiarity of being on the summit of Lookout Mountain.

As we find so often in other coal-fields, one prominent seam pervades the whole area, thus the Sewanee seam is found in the main Tennessee coal-region to furnish a larger amount of coal than any other single seam in the State. This coal has excellent properties, though it varies somewhat in its characteristics in different localities. It is deemed a good coking coal, and is effective in producing steam; it is almost free from sulphur, and is excellent for domestic use. The largest coal-mining operations in the State are in connection with this seam. The mines are within the Little Sequatchie Valley, and the vicinity of Tracy City may be taken as the main site of these mining operations, while on the head-waters of the Sequatchie, in the north-eastern portion of the valley, are many coal-seams that are worked.

Alabama Coal-Field.—In its northern portion, amid the mountains and highlands of the southern extremities of the Alleghanies, Alabama is rich in mines of coal and iron, lying almost side by side. The area of the coal-fields of the State may be safely estimated at 5,500 square miles. The famous Black Warrior field, thus named from the river that runs over its surface, has much the larger portion of this coal—about 5,000 square miles.

The Cahawba field is long and narrow, and the Coosa field—both named from the rivers in whose valleys are the coal-seams—make up the remainder of the coal area. The coal of the Black Warrior basin

is represented as equal in heating power to the similar varieties of coal in the more northern portions of
the Alleghany bituminous field: some are dry burning, others coking, having about 65 per cent of fixed
carbon, while others contain more volatile matter.
The fixed carbon in the coals of these several fields
range from 56 to 66 per cent, while the seams range
in thickness from three to seven feet; that between
the Coosa and the Black Warrior, in the vicinity of
Birmingham, is nearly five feet.

Here is a remarkable instance of coal, iron-ore,
and limestone, being found near one another in a
native state. Beds of iron-ore of the red or brown
hematite variety abound in Red Mountain — thus
named from its red soil, caused by oxide of iron.
This range is twenty-five miles long and a number
of miles wide; on both sides of it are coal-fields, as
if the long mountain had been pushed up in the
midst of the latter. In a central position, between
the Black Warrior and the Coosa Rivers, amid these
coal-fields and deposits of iron-ore, has been founded,
since the close of the civil war, the city of Birmingham, which is making rapid strides toward becoming a center of the industries pertaining to the manufacture of iron. It is situated in a beautiful valley,
eighty miles long and ten wide, and which is penetrated by a number of railways.

Georgia Coal-Field.—In almost immediate connection with the northeastern coal-field of Alabama
may be mentioned the limited field located in North-

western Georgia; the latter's area is only about two hundred square miles. There are in this field, belonging to the region of Lookout Mountain, five seams of coal. These vary much in their thickness, ranging from one to five feet. For some unexplained reason the coal of this field is not of the best quality.

Ohio Coal-Field.—We come now to treat of that portion of the Alleghany coal-field that lies in Northeastern Ohio, and to some extent in Northwestern Pennsylvania. All the streams of this coal-basin take a south or southeastern course toward the Ohio River, as we have seen those of the other fields, and with only one exception flowing to the same river. The latter flows northerly into Lake Erie. The field within the State of Ohio is about 180 miles in length by eighty in the widest part, and has an area of workable coal of at least 10,000 square miles. This coal-region lies on the northwest side of the Ohio River, whose general course it follows in a southwest direction from the vicinity of Youngstown to Ironton; some miles from the latter town, the river, having changed its course, commences to flow toward the northwest across the coal-field; meanwhile, in cutting its channel, it carried away the coal that impeded its progress.

The Ohio coal-seams are very regular, and have a gentle inclination; the strata appear never to have been disturbed by internal convulsions. There are three classes of coal in this field. The dry, open-burning, or furnace-coals (block), coals that coke, and cannel — the latter extending across the line into Pennsylvania. The most important portion of this field is in the northeastern part, in the vicinity of Youngstown, together with the Mahoning Valley. The main seam of the latter is the most valuable in the State; it is of workable thickness, very pure, and well adapted, even in a raw state, for smelting iron-ores. This coal is also very compact, and comes out in large blocks; hence it is called by the miners "block-coal" (p. 66). In the Mahoning Valley is also a vast deposit of iron-ore. These two advantages combined have made this the most important region for the manufacture of iron within the State. The Hocking River Valley coal-seam underlies an area of six hundred square miles, and has a thickness ranging from six to eleven feet, while it is remarkably uniform in its structure and pure in its composition. The cannel coals of this State have a large percentage of hydrogen, and the gas obtained from them has in consequence a highly illuminating power. The presence of this large proportion of hydrogen is, in theory, attributed to the process by which these coals were formed, the deposit being in lagoons of open water in the coal-marshes, on the bottoms of which was accumulated the carbon of the vegetable tissue steeped or macerated in the water, together with a large amount of the remains of all kinds of fishes and other aquatic animals.

6

VII.

THIS field occupies an area of 50,000 square miles, the coal underlying parts of three States, in the following proportions: 35,000 in Illinois, 10,000 in Indiana, and 5,000 in Western Kentucky. The latter is separated from the two former by the Ohio River, while the Wabash is the dividing line between the fields of the first two States. The coal of the central field is bituminous, but its good qualities, upon the whole, are not quite equal to those of the same class of coals in the Alleghany field, the latter having, for the most part, more fixed carbon and less water. To this statement there is one great exception.

Block-Coal.—A deposit of coal, having peculiar characteristics, was accidentally discovered (1869), when sinking a well at Brazil, in the State of Indiana, then an obscure station on the Indianapolis and Terre Haute Railway, about sixteen miles from the latter place. The coal lies eighty feet below the surface of the prairie. This seam has since been traced, with few interruptions, from about one hundred miles south of Lake Michigan to near the Ohio

River, a distance of more than one hundred and fifty miles. Deposits of block-coal have since been discovered in many places in Indiana and also in Illinois, and numerous mines have been opened in both States. The block-coal area around Brazil is about four hundred and fifty square miles, and the total depth, including all the seams, is twenty-eight feet; some of these, however, are too thin to be workable, but the one available is six feet thick.

It is named "block-coal," by the miners, from the formation of the coal itself, which in the mine is in cubic blocks, two or three feet long, and a foot or more wide. No one can explain its peculiar formation nor its perfect regularity. It is a mineral charcoal slightly connected with bitumen, and so pure, for the most part, that when handled it scarcely soils. It burns with a bright-yellowish flame and crackling noise, is quite free from sulphur, and does not clinker; the block when burning retains its shape till reduced to ashes, the latter being as white and flocculent as those of hickory-wood. When analyzed it gives from 57 to 62 per cent of fixed carbon, and a small amount of water and of ash, while there is scarcely a trace of sulphur. The structure is laminated or in layers, in one direction splitting easily, but not in the other. In smelting iron-ore the blocks retain their shape to such an extent that the blast and flame find an easy passage through the entire mass of fuel, ore, and flux. Thus, for making iron direct from the ore this raw coal is not surpassed by any other, and yet its efficiency is greatly enhanced by being mixed with the famed coke of Connellsville.

Block-Coal utilized.—We have already noticed that the materials for making iron are often near one another, but at Brazil there are only two in proximity—limestone and coal; the ore has to be obtained elsewhere. Experiment shows that the mixture of iron-ores from Lake Superior and from Iron Mountain, Missouri, can be used with the best results to produce steel by the Bessemer process. When the good properties of the block-coal became known, enterprising gentlemen determined to utilize it in making iron and steel. It was found easier to bring the ores from Lake Superior and from Iron Mountain to the coal, rather than seek some central position to which both the ores and the fuel must be brought. A number of railways have extended their lines so as to reach this coal-field, or to connect with others that do. There are in operation at Brazil some half-dozen or more furnaces for making iron, and this hitherto obscure railroad-station has become a center for that manufacture. The block-coal is carried in immense quantities to numerous iron furnaces and foundries in the West, and north to the works on the lakes for smelting copper as well as iron. Specimens of coal taken from twelve separate localities in the State, when analyzed, range from 46 to 63 per cent of fixed carbon, while the average of the whole is 53.

Illinois Coal-Field.—In this State the discovery of coal was made more than two hundred years ago by Father Hennepin, a French missionary to the Indians. He noticed the outcrop near where now stands the city of Ottawa, and recorded the fact in his journal, which was afterward published, and is said to be the earliest notice in print of the existence of coal within the boundaries of the United States.

The coal of this large field is unfortunately more or less injured by the presence of sulphur, except the portion that belongs to the block variety. To free it from this injurious compound is a very difficult operation; it can be done only by coking, and that often fails.

Illinois Block-Coal.—The most valuable coal district of the State is that of the iron-smelting block-coal in the vicinity of Carbondale, in Jackson County, of which about one half is underlaid with block-coal. The latter partakes of the qualities of the same variety found at Brazil, in Indiana. This coal deposit has the advantage of being within easy distance of St. Louis, and almost directly opposite the immense deposits of iron-ore found in the famous Iron Mountain in Missouri. This pure coal is peculiar in being in two seams, one of three feet in thickness, the other of two, while between them is a partition of a few inches or more of shale, which is easily removed. It is hard and bright, and the layers are separated by a sort of mineral charcoal. The coal itself has 66.5 per cent carbon, and volatile matter 25; the remainder moisture and ashes—the latter being only about 3 per cent.

Geologists say there are altogether *sixteen* different seams of coal within the boundaries of this State. These are never all found in one locality, neither are they uniform in thickness nor in quality; some extend over large areas, and are so available in thickness as to be easily worked, and, not being far below the surface of the prairie, are reached by means of shafts. The latter varies in depth from about one hundred feet and upward—904 being the greatest thus far reached. The thickness of the workable seams also varies between the extremes of two and a half feet and ten. Between these extremes, the seams are, generally, from four and a half to seven feet in thickness, and upon the whole the coal is easily mined, the seams being horizontal and parallel to one another, and so extensive as to be in abundance for all purposes, while the presumption is that there are still more areas to be discovered within the borders of the State.

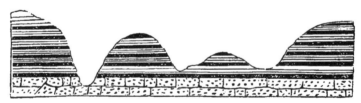

FIG. 9.—Illinois Coal-Field (after Daddow).

Western Kentucky Coal-Field.—It has been remarked that the coal under the prairies is reached by shafts ranging from eighty to four hundred feet in depth, but south of the Ohio it is different. The Western Kentucky field belongs to the great inclined plane previously mentioned (page 49), and this accounts for the seams of coal being generally above the streams and in the hills; the latter in some respects are quite remarkable. In one instance near the village of Providence, in one hill-side within the space of one hundred and twenty-five feet, are found three seams of different kinds of coal, each from five to six feet in thickness. In this field are seams of valuable cannel coal, as well as the ordinary bituminous. They border on the Ohio, and extend southward up the rivers and valleys. Over this entire coal area there are at least two workable seams; the latter are easy of access, and afford great facilities for mining. Says one authority: " In this continuous coal-bank, in the heart of a fertile country, is a mineral wealth of more real value than the gold of California. Here is a series of high hills, which from the base to the top look like a succession of coal, iron-ore, and limestone strata, heaped there as an inducement to labor, capital, and enterprise."

Michigan Coal - Field.—This deposit of coal is located in the central portion of the State, and between the Lakes Huron and Michigan. The seams are few, only two are definitely known, and, when compared with the fields elsewhere, are thin. Of these one pervades the whole formation; it ranges in depth from three feet to five. The coal of this field in quality is very bituminous; when burning it blazes with a bright and strong flame, but is deficient in heat-producing properties, as it contains only 45 per cent of carbon to 49 of volatile matter, and ash 2 per cent, the remainder being sulphur and water. The extent of the workable coal area of this State is about 7,000 square miles. " It belongs to the great Carboniferous formation of the Alleghany field," and the shales are rich in the remains of terrestrial vegetation, such as specimens of fern-leaves in a beautiful state of preservation.

Iowa Coal-Field.—The coal area immediately west of and bordering on the Mississippi evidently belonged originally to the central field, from which it was separated by the Father of Waters. There is no very essential difference between the properties of the coals of the two fields. That of Iowa contains from 45 to 50 per cent of fixed carbon, and from 35 to 40 per cent of volatile bituminous matter. The great body of the coal area of this State—16,000 square miles—is in the south and in the middle of the State, commencing north of Fort Dodge, and following the valley of the Des Moines River. The seam worked at the former place is from five to six feet thick; this, perhaps, is the most important portion of the coal area of the State. There appear to be at least two workable seams pervading the State, but as to their thickness they are quite irregular, the lower one being the most reliable as to quality, as it contains less sulphur. Toward the northern portion of the State the coal-measures disappear almost entirely. Limestone is here found everywhere in connection with the coal. " Much of the Iowa coal is evidently of an inferior quality, owing to the presence of so large a quantity of sulphur, moisture, and other impurities." (*Coal Regions, etc., p. 461.*)

Missouri Coal-Field. — The bituminous coal of Missouri has properties similar to that of Iowa, and it extends from the southern line of the latter State southward to beyond the latitude of Sedalia and Jefferson City, trending to the southwest, but northwest of the Osage River. The Missouri River runs across this field, dividing it into two unequal parts. The coal of Missouri has the general characteristics of that of Iowa, and is also reached by shafts, the depth of which depends upon the undulations of the surface, as the seams are horizontal, or with very little dip. The depths of the shafts are sometimes as low as twenty-two feet, but often down to forty. The coal-seams reached vary in thickness from two to five feet. The entire coal area of the State is, according to Prof. Swallow, State Geologist, 27,000 square miles. The coal from a prominent seam was proved to contain 53 per cent of fixed carbon.

We leave these regular seams, to speak of a unique class of coal deposits. These singular beds are limited in extent, but are numerous and unusually deep or thick, and of course do not partake of the regularity that is found in seams. They are located in what appear to have been ravines or deep valleys, and geology says amid the older rocks, below the level of the regular seams of coal. These deposits furnish cannel of the common bituminous variety, and some of the best coal in the State for making gas, and producing an abundance of flame. These immense beds are in thickness from twenty to forty feet; they are found near the mouth of the Osage River, and at the same point on both sides of the Missouri. This coal has a little more than 51 per cent of carbon, and nearly 42 per cent of volatile matter, and it is so light that before it becomes saturated it will float in water. To account for these singular deposits of coal, geologists have various theories. One is, that while the coal material was in the form of peat, and in a semi-fluid state, when upheavals came that made slopes of portions of the bottom of the peat thus saturated, and it sliding down these inclined planes, accumulated in masses within these lower places, and afterward, by the action of heat and superincumbent pressure, the whole mass was transformed into coal.

Kansas Coal-Field.—This is a continuation of that of Iowa and of Missouri. Kansas has two coal-fields, the carboniferous and the lignite. The former extends across the eastern portion of the State into the Indian Territory, being in length 208 miles north and south, and east and west on an average 107; thus the east end of the State has 22,256 square miles of coal. It lies in nearly a horizontal plane, having undergone but little change from its original position. The coal is reached by means of shafts from any point on the surface of the prairie. Amid these coal-measures, and partially coextensive with them, are ledges of limestone.

There are two important seams of coal in the State belonging to what geologists term the lower measures. These seams are respectively three and six feet in thickness, while the upper ones are much thinner, so that they are seldom available for working. These two seams mentioned are a little more than one hundred feet apart, the lower one being better as to the quality of the coal and also the thicker. The latter is about three hundred feet below the surface of the prairie. The deepest shaft in the State is near Leavenworth; the coal is 712 feet below the surface. The greater portion of the coal found in Kansas is good; it is usually free from sulphur and other impurities; has a bright appearance when broken, and does not crumble to dust by handling in the transporting; it also burns with the clear white flame of Pittsburg coal, and owing to its heating properties is good for domestic purposes, and is universally used for the same reason by the blacksmiths in welding iron.

Nebraska Coal-Field.—This contains only about 3,600 square miles of coal area thus far discovered; its seams are generally quite thin, though the coal partakes of the properties of that found in the States adjoining. In boring for coal, a seam of two feet and a half was found 820 feet below the surface. This field appears to be on the western margin of the great Carboniferous coal-basin, within whose limits were prepared such immense treasures of fuel.

Arkansas Coal-Field.—Geologically speaking, all the coal deposits of this State belong to the lowest member of the coal formations; not a trace of coal has been found on or near the tops of her mountains or highest hills. Two seams of coal have been discovered, but only the lower one is of workable thickness. The best seam, however, yet found in the State, is near the mouth of Spadra Creek, a tributary of the Arkansas. It is a semi-anthracite or semi-bituminous, and quite as rich in carbon as some of the minor coals of the Pennsylvania anthracite region; the average percentage of carbon, as ascertained from six mines, being 84. That is richer—if the analysis is correct—in carbon than some of the semi-bituminous coals of Pennsylvania and Maryland. It is pronounced, as far as known, to be better than any other Western coal for manufacturing purposes, where intense heat and durability are required. The seams range in thickness from four feet to seven. They crop out along the river, while there are indications that this coal extends far and wide in the vicinity.

It is a puzzle of geology to account for this deposit of semi-bituminous coal, as the formation is level and undisturbed, and gives little evidence of the heat and pressure requisite to transform the original mass into that class of coal. But it may be suggested that as the Hot Springs are distant not quite one hundred miles, why may not the heat, that still lingers there underneath the surface of the earth, have extended at one time as far as this coal-field, and thus driven off the volatile matter and transformed the original mass into semi-bituminous coal? The coal area of Arkansas, thus far discovered, is about 11,000 square miles.

The Indian Territory has a coal area of about 12,000 square miles, but its fixed carbon is quite low, being only about 36 per cent.

Texas Coal-Field.—In the northwestern part of this State bituminous coal is found on the banks of the Brazos River, in the vicinity of Fort Belknap. The seams lie horizontally, and are from two to five feet thick. The characteristic fossil forms of the Carboniferous age are found in connection with this coal, and the fossils obtained from the carboniferous limestone indicate that the vegetable ingredients in the coal are a product of that age. The coal is of fair quality, burns freely, and leaves a white or gray ash; in addition, another and lower seam, sixty feet beneath the surface, was discovered in sinking a well. The coal of the latter can be taken out in large blocks, but it crumbles somewhat when exposed to the atmosphere. The area of this special field is estimated at 5,000 square miles, while that of the entire State is estimated at 25,000.

The Texan coal-measures, for the most part, have the characteristic carboniferous fossils of the central field, and the coal is similar in its general properties.

It is of different grades of excellence; while in one district the coal contains 53 per cent of carbon and 36 of volatile matter, another is reported as having 61 per cent of carbon and 36 of volatile matter, and one seam near the town of Washington, on the lower Brazos, claims to have 75 per cent of carbon and 22 per cent of volatile matter, and 3 per cent of ashes. The latter grades are equal to the best qualities of the bituminous variety. It would appear that the greater portion, if not the entire valley, of the Brazos was underlaid with coal. In the southwest portion of the State, west of the ninety-eighth meridian, is an extensive field of excellent lignite coal. It evidently underlies the upper branches of the Nueces, and "extends along the Rio Grande from Laredo to above Eagle Pass," at which place it crosses the river into Mexico. This coal supplies the railways in that section, and is shipped on them to the adjacent cities. A fine deposit of coal has been recently discovered on the Colorado River—the seam four feet thick.

A bituminous coal-field has been discovered in Jack County, in the northern portion of the State. The indications are that the field is extensive, and the coal of fair quality. The seam is about four and a half feet in thickness.

The Extent of Pure Carboniferous Coal.—We here close our survey of these interesting coal-fields, the coal of which is the product of the *same Carboniferous age*, but, owing to certain influences in the economy of Nature in different portions of the area, the coal assumed several characteristic phases. Hence we have the bituminous, the original form or base, the cannel, the semi-bituminous, and the anthracite. In heating power and value for the purpose of domestic use or for manufacturing in every form, the coal of this Carboniferous age excels all others. Its area extends from the eastern margin of the anthracite of Pennsylvania, except a few intervals, to Western Iowa. It sweeps along the eastern foot-hills of the Alleghany Mountains, taking in the semi-bituminous, and on their western slope the famous field known as the Alleghany; it includes the central field, north and south of the Ohio; crosses the Mississippi, and from Northern Iowa passes south over three States; then diverging to the southwest, underlies portions of Arkansas, of the Indian Territory, and of East-Middle Texas. · In all, this coal area of the Carboniferous age amounts in round numbers to 205,000 square miles. Everywhere the coal contains a number of workable seams one above another, and the aggregate depth of these, when taken together, is fairly estimated at *twenty feet* of available coal.

VIII.

IN the last section we closed our review of the immense coal area of the United States, which was the product of the Carboniferous age, and in which coal the carbon so predominates as to impart to it an intense heating power. We come now to treat of the product of a later Carboniferous age—Cretaceous (chalk) or Tertiary, according to geology— in the coal of which carbon does not predominate so much, but bitumen and gaseous properties more, constituting a variety of coal of less heating power, and known as lignite. The transition from the bituminous to the lignite variety commences in Western Kansas near where the 97th meridian line intersects the 39th parallel of north latitude, which point of intersection closely marks the territorial center of the United States, not including Alaska. This meridian line, thus marking the transition with but little variation, extends from the southern border of the United States, on the Rio Grande, to the extreme north in the British possessions, for the lignite coal has been traced along the eastern slope of the Rocky Mountains, and down the banks of the Mackenzie River, while to the east of that line bituminous coal, the product of the Carboniferous age, is found on Melville Island, in the Arctic Archipelago.

The Marsh-Basin; The Plain.—In theory we can imagine that long before the era of the Alleghanies and other upheavals, the portion of the territory of the United States east of that meridian was for the most part a vast marsh or basin, covered in the Carboniferous age by a rank vegetation suited to a moist climate, such as the many varieties of ferns, and by a dense undergrowth; and in a later period, west of the same line, where now stand the Rocky Mountains and highlands east and west, was a territory that was undulatory on the surface, and covered by dense forests of deciduous trees which, from their nature, could not flourish in a climate so moist as that of the Carboniferous age. That the lignite was formed from the product of trees, is inferred from the impressions of their leaves imbedded in the clay beneath the coal. These specimens are easily recognized by their veins, as well as small branches and the woody fiber, looking like charred wood or soft charcoal. Here are found representatives of the ordinary forest-trees of to-day, such as oaks, walnuts, poplars, beeches, sycamores, elms, lindens, and others long since extinct in that region. Says Prof. Hayden, of the United States Survey, "They" (these specimens) "are most perfectly preserved, and all plainly point to a period far back in the geological past, when these vast treeless regions of the present time were covered with dense forests."

Ferns and Carbon.—There being no trace of ferns in the lignite variety, is an evidence that the bituminous coals, in whose formation the ferns so much predominate, owe their greater heating power to the carbon thus imparted to them. Though the heating power of lignites is scarcely sufficient to prepare iron for welding, yet its power to produce flame makes it an excellent fuel for generating steam, as the flame passes easily through the tubes within the boilers, and thus makes up the deficiency of heat by diffusing it over a greater surface. Analyses of seven specimens of western lignites collected from as many far-separated localities, and under the superintendence of Prof. J. S. Newberry, showed an average of 63 per cent of carbon. Though these . lignites are near at hand, the repairing or machine shops of the Union Pacific Railroad use Eastern semi-bituminous coals, because of their greater production of heat.

Characteristics of Lignite.—The deposits of lignite are not in regular seams, like the coal of the Carboniferous age, but in isolated beds of varied thicknesses. Instead of being formed, as the bituminous, from steeped or macerated vegetable matter, the lignite appears to be composed of pure woody fiber, and is defined as a " mineral coal retaining the texture of the wood from which it was formed." Hence the theory has been entertained that this woody material, trunks, branches, and leaves of trees, was in various ways collected in great masses—some in heaps of fallen timber within morasses; some carried by the force of water into ravines and piled up, or have been lodged in the channels of rivers—and that in the course of geological ages these deposits were transformed into lignite.

Extent of Lignite.—The area of lignite coals within the United States, lying between the meridian line of 97 and the Rocky Mountains, is estimated at 50,000 square miles. The deposits being more or less isolated from one another, this estimate is designed to include the actual workable beds of the coal, and not merely the territory where it is known to exist, for the latter would equal more than 300,000 square miles. " Scientifically speaking, there is no more coal west of Kansas, thereby meaning that better quality of fossil fuel found among the rocks of the Carboniferous age, but, practically and commercially, there is a great abundance of lignite. During the Cretaceous and early Tertiary periods an immense swamp seems to have existed all along the foot of the Rocky Mountains, on the eastern side extending from New Mexico northward for many hundreds of miles into the British possessions." (*Coal Regions, etc., pp. 534, 535.*)

Dakota and Montana Lignites.—Lignite coal appears to be diffused over the entire region between the 97th meridian and the Rocky Mountains, as the settlers find it along the Northern Pacific Railway. "Seams of lignite abound, cropping out from the hill-sides of Western Dakota, and the settlers obtain their supplies for domestic use at no greater cost than getting it out from the faces of these out-crops. This coal has about three fourths the heat-producing capacity of ordinary bituminous." In the vicinity of Coal Harbor, North Dakota, on the Mis-souri River, has been recently discovered a deposit of lignite coal. The seam is unusually thick, being, it is said, eighteen feet, while the indications are that the area thus underlaid is quite extensive. "Coal in abundance is found in Montana near the Yellowstone, Musselshell, and Missouri Rivers, in immense beds of lignite which crop out along the bluffs. Veins of a harder coal, that can be coked has been found also in the Belt Mountains." (*Hist. Northern Pacific Railroad, pp. 337–340.*) A large deposit of coal is also reported as having been found in the vicinity of Fort Benton. This coal is stated to be susceptible of coking.

Wyoming Territory.—How often, in the order of Providence, is it seen that treasures lie in the earth undiscovered until man is prepared to utilize them! A striking incident, in illustration of this principle, occurred in the discovery of coal in this Territory. The lack of fuel for the locomotives was earnestly urged against the building of the Union Pacific

Railway, because of the extra expense in operating it, there being neither wood nor coal along the route. However, the road was nearly finished, and the company were making preparations to supply fuel, and a contract was made with a firm in Pittsburg to furnish a certain class of cars to transport the coal to relay-stations along the road. The necessary papers were drawn, and on the following day the respective parties were to meet at a certain time and place to sign the contract. The parties met; but, only about an hour previous to the meeting, a member of the corporation received a telegram, announcing that coal had been discovered at Cheyenne, near the railroad. The contract was not signed. In sinking a well the workmen found indications of coal; the clew was followed, which resulted in finding a bed of lignite *seven* feet thick and at *seventy* feet beneath the surface. Coal was now sought for and found at various places along and near the road for 500 miles west of Cheyenne and about 200 east; from these numerous mines the railway derives its fuel.

Here commenced the development of the coalfield of Wyoming, which is estimated by competent authority to be at least 20,000 square miles, the whole being remarkable for the thickness of the beds. The field lies within the southern half of the Territory, and occupies portions of the valleys of the North Platte and Sweetwater Rivers. In what is known as Coal Ridge in the northern portion of the

field there are extensive deposits which are remarkable for their depth, three being reported respectively as forty, sixteen, and twelve feet in thickness.
(See Fig. 10, which illustrates the dip and the seams
of this coal-field of Wyoming.) The spaces between
the seams of coal are filled for the most part by sandstone and beds of fire-clay. " All of these beds
(thirty-four in number) are confined to about 2,000
feet of strata."

According to the report of Gov. Warren to the
Secretary of the Interior (1885), these coals range in
heating power or fixed carbon from 51 to 53 per cent.
One belt of coal extends for 120 miles, and Prof.
Aughey, Territorial Geologist, is quoted by the Gov-
cruor as saying that "along this belt there are from
three to eleven coal-beds, varying in thickness from
twelve to twenty feet." The strike of these beds is
nearly due north, and they dip westerly from 20° to 23°.

Says Prof. Hayden, United States Geologist:
" The coal of the Rocky Mountains is distributed
along their flanks as several leaves in the great book
of folded strata. Nowhere in the world is there
such a vast development of the recent (Tertiary or
Cretaceous) measures, and in few places is their existence more necessary to the advancement and improvement of the region in which they occur. They
lie regularly and in the main quite horizontally ; being close to the mountains, the beds are naturally
tilted." The coals of this section, though lignites,
have many of the characteristics of the bituminous,

but they are not so uniform in their character as the latter, and differ from one another more in different localities than does the bituminous.

Idaho Coal. — In Central Idaho, about thirty miles west of Salmon City, a vein of asphalt was discovered in 1885, near the mining town of Dynamo. It is in one of the foot-hills of the Salmon River Mountains, and ranges in thickness from one and a half to three and a half feet, and has a dip

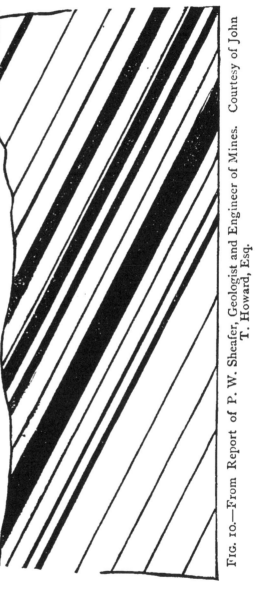

FIG. 10.—From Report of P. W. Sheafer, Geologist and Engineer of Mines. Courtesy of John T. Howard, Esq.

of about 60° as it rises from the edge of the creek or valley to 160 feet in the hill. The extent of the deposit is not fully known.

Colorado Coal-Field.—In this State coal or lig-nite has been found on both sides of the South Platte River, and farther south, in the vicinity of Golden City, are several isolated beds nearly vertical in their position. One of these ranges in thickness from ten to fourteen feet at the outcrop; the bed proved to be, however, very irregular in depth, run-ning down to even a few inches, then expanding to eight or ten feet, and on the whole averages perhaps *five* feet. A number of beds have been opened on Boulder Creek, twenty-two miles north of Denver; among these is a bed of coal nearly *nine* feet in thick-ness and of a superior quality in respect to its heat-ing power, so that it can be used in welding iron.

In the west-middle portion of the State, in Gar-field County, is a remarkably valuable coal-field " ex-tending from Grand River near the mouth of Elk Creek in a southeastern direction for thirty miles." One seam of this coal is stated to be " thirty feet in width, extending above the water-level in some places over 1,200 feet to the crest of the mountain above." Within the space of about 100 feet are three other seams of coal, varying from five to eight feet in thickness. These seams are almost vertical, and therein consists the peculiarity of their position, as they extend from the top of the mountain to an indefinite distance below the water-level. This coal is susceptible of being coked, and thus made avail-able for the smelters of silver and lead ores found in the vicinity.

There are numerous other deposits of coal found within the State, and no doubt many more will yet be discovered, and sufficient to supply its domestic wants. Analysis shows that the percentage of fixed carbon, in two specimens from representative mines, ranged from 41 to 42 per cent. Anthracite is said to have been also found in this State. The coal area of the State has been estimated from 20,000 to 50,000 square miles. (*Min. Resources U. S., p. 24.*)

Utah Territory.—Near the line of Wyoming and Utah Territories, at Evanston, 441 miles west of Cheyenne, are found a number of valuable coal-beds. One of these is remarkable for its thickness, that being twenty-six or twenty-seven feet; but only from eight or ten feet of the coal is taken out, as more convenient. The coal, however, is not a favorite with the locomotive-engineers, because of its clinkering and of not burning up cleanly. Numerous other mines are found in the hills of Utah, sufficient, it is supposed, for domestic use, the coal having the characteristics of the lignite variety. There is a marked peculiarity belonging to some of the lignites, that of taking fire spontaneously. This trait was noticed in 1804, by the first American explorers of that region—Clark and Lewis—on the Shoshonee (vulgarly, *Snake*) River. Mines are to-day on fire from this cause in many localities within our Territories, as well as northward along the Mackenzie River.

New Mexico Coal-Field.—In this Territory are

two prominent deposits of coal belonging to the same general line of beds that we have noticed along the eastern foot-hills or base of the Rocky Mountains. They are both within the valley of the Pecos River, the solitary branch of the Rio Grande. One, the Corrillas, lies higher up the valley, and the other, White Oaks, farther down, but both are in a southeast direction from Sante Fé, the capital of the Territory. It is singular that of these two fields the former should be of anthracite and the latter of bituminous, and yet so near each other. Why this peculiarity, is left for geology to explain.

Prof. Raymond, in his report, 1870, on the " Mineral Resources of the United States," speaking of this anthracite, says: " As far as its application for all practical purposes is concerned, it is fully equal [?] to Pennsylvania anthracite, and is really the best fuel ever discovered, so far, in the West." Prof. J. S. Newberry characterizes this as *lignite anthracite,* and, by analysis under his direction, it is found to contain 74 per cent of carbon ; Pennsylvania anthracite has 95 per cent. The bituminous seams are seventeen in number, ranging in thickness from one to six feet. Southwest of these mines at Los Bronces, in Northern Sonora, Mexico, are mines of lignite anthracite, which, according to the same analysis, contains 84 per cent. (*Coal Regions, etc., p. 557.*)

Arizona Coal-Field. — This field extends north and south in detached portions from near the Gila River, on the south and on the eastern slopes of the

Apache and Mongollon Mountains, up into Southern Utah, meanwhile sending a branch east into North-western New Mexico. These beds are also on the head and middle streams of the Little Colorado River, as well as on some of the tributaries of the Gila. The lignite coals are not so uniform and con-tinuous in their seams as those of the bituminous of the Carboniferous age ; hence we find the deposits of the former more detached from one another. There appear to be several deposits of coal in Arizona. In the south, the one in the valley of the Gila, on Deer Creek, one of its tributaries, the coal is of ex-cellent quality and will coke. The seams are from three to eight feet in thickness, and the deposit is about four miles long by two wide. North of this is another and still more extensive deposit in the vicin-ity of Fort Defiance ; here the seam is nine feet thick : this bed also extends northeast and east into Northwest New Mexico. Then comes another field in a northwest direction from the latter, but which extends north into Utah ; here the seam or bed is twenty-three feet thick.

The coal of Arizona has been described by Mr. C. P. Stanton, " a competent geologist," as possess-ing " all the qualities of bituminous coal and to rank next to anthracite for domestic purposes," saying: " I see no reason why it should not be pre-eminently useful for generating steam and for smelting ores. . . . This description will apply to all the coal in the great Arizona coal-basin." (*Hamilton's Arizona, p. 243.*)

IX

California Coal-Fields.—The coal on the Pacific slope partakes of the same general character of the lignites on the east side of the Rocky Mountains. Twenty-eight miles nearly east of San Francisco are, as far as known, the most important coal deposits in the State of California. These consist of coal-seams in a ridge of hills on the north and northeast sides of Mount Diablo, a coal district that belongs to the Coast Range. The outcrop extends for ten or twelve miles in an east and west direction, and in these seams the coal is found— the lower one being four feet in thickness, the middle two and a half, and the upper three feet eight inches.

There are some half-dozen mines in operation in taking coal from these seams; the lower one is worked by a shaft 700 feet deep. San Francisco for the most part derives her coal-fuel from these mines, for domestic use as well as for manufacturing and the driving of steamers and locomotives; they also supply the cities around the bay and up the valleys of the rivers. The coals of the Pacific slope, being of the later geological formation, are soft and friable, and deficient in heat-producing qualities, when compared with the bituminous coal of the Alleghany field. The average amount of fixed carbon in the coal taken from five different mines at Mount Diablo, by analysis, is 43 per cent, while that of the Pittsburg seam is nearly 55.

Coal is also found within the State in many other places, in the Coast Range as well as amid the western foot-hills of the Sierra Nevada; but the quality is not generally of the best, while the quantity is quite limited. In a southeast direction from the Diablo coal-field, coal occurs in the valley of the San Joaquin, and also in the southern extension of the Coast Range. The coal found in the latter is not as valuable as that obtained from the mines in the vicinity of Mount Diablo; that in Ione Valley, Amador County, has some marked peculiarities. The deposit is for the most part in a narrow trough of varying width and stretching for more than twenty miles, and at the general depth of sixty feet below the surface. The seam varies in thickness from four feet to twenty. Still farther southeast coal is found in Los Angeles and Mono Counties, and also in that of San Diego. The latter appears to be the most valuable, as the seam is solid and ranges from four to seven feet in thickness. The coal is mixed a little with slate, but in the main differs scarcely in its characteristics from that of Mount Diablo. This portion of the State is almost destitute of timber, causing the coal to be of unusual value to the people.

The Coos Bay Coal-Field.—This bay penetrates Southern Oregon for fourteen miles, and in the upper portion curves sharply toward the north; it varies in width from one and a half to two and a half miles. Coos River enters it at the northern end; and on that river, four miles from its mouth, is the coal-mining district, in the midst of a hilly and densely wooded country. The coal lies in three seams near one another. The two lower are peculiar in their position, as between them is a vein of sandstone *four* inches thick, while the seams of coal are each two feet three inches, but the whole is mined and the sandstone is picked out. The upper seam is of inferior quality, and, being only one foot thick, is left to serve as the roof of the mine. This coal is free from sulphur, and is not liable to take fire spontaneously.

Upon the whole, this coal-field appears to be unusually large for that region and regular in its conformation, as it is known to extend for eighteen or twenty miles south of the bay, and besides for many miles inland. The main seam varies in thickness; in some localities it reaches even nine feet. When freshly mined the coal in appearance resembles that of Illinois and Iowa, though it differs in being lignite, as in it can be plainly seen the structure of the wood from which it is formed.

" Masses of several hundred pounds' weight are seen, which were evidently portions of the carbon-ized trunks of trees of large size, in which the rings of annual growth, knots and branches, were almost as plainly perceptible as in recent wood." Not-withstanding this apparent solidity, when exposed for a time to the air, it cracks and breaks into numberless cubical fragments. It burns freely, with a bright, cheerful blaze, and gives out a fair pro-portion of heat; it produces gas largely, but of an inferior illuminating power. It contains nearly 47 per cent of carbon and about 50 of volatile matter.

Seattle Coal-Field.—In Washington Territory we find coal-mines near Seattle. This town is on the east side of Puget Sound, and has the finest harbor belonging to that magnificent inland sea. The coal-field is nine miles distant. In this field are five seams of coal, ranging in thickness from *four* to *twelve* feet, and covering a large district of many square miles. The coal contains 46 per cent of carbon and 35 of volatile matter, 12 of water, 6 of ashes, and 5 per cent of sulphur.

Bellingham Bay.—Directly north of Seattle, on Bellingham Bay—which is fourteen miles long and three miles wide—is deemed the largest and finest coal deposit in the Territory. The main seam is *nine* feet thick, all of which is available for mining; the coal being of the same general quality of the others on the coast, having of fixed carbon 48 per cent and of bitumen 50. The mouth of this mine is near the water of the sound, where the largest vessels can easily come up to the wharf. There are several seams; the coal when burning often leaves an unusual amount of slag and cinders. "The flora of Bellingham Bay is remarkably like that of the lignite-beds of the upper Missouri, the *genera* being all represented and some of the species identical" (Prof. Newberry). Similar specimens of flora are found in the coal-mines on Vanconver's Island.

The Coast Outcrop.—South of Cape Flattery twenty-five miles is an extensive deposit of coal very similar to that of Coos Bay. In addition there is almost a continuous outcrop of coal along the coast-hills toward the south as far as Santa Clara southeast of San Francisco, where is found a deposit of lignite having the characteristics of that of Coos Bay, but said to be somewhat inferior in its properties. "Farther inland there is probably an extensive region underlaid with coal on the west side of the Cascade Range, between Willamette Valley and Bellingham Bay—the lowest seam being reported to be *sixteen* feet in thickness."

Alaska Coal-Field.—Coal is found in numerous places in this Territory, though the seams appear to be rather small. It has been used in some instances with success by the United States steamers when cruising in the adjacent waters. It is found not only along the south coast, but in the interior; but at " Cape Beaufort, on the Arctic coast, is a small seam of true Carboniferous coal." Can the latter have any connection with the coal of the same geological age on Melville Island? (p. 17). The coal from the deposit on Cook's Inlet has of fixed carbon 50 per cent and of volatile matter 40. Anthracite is also said to be found in Alaska in various places, but in limited quantities. It may be noted that the coal of the Cretaceous or Tertiary period—the lignite—is found along the east coast of the Pacific, the seams at intervals cropping out of almost every bluff from Sitka, Alaska, as far as Chili—where lignite is found —if not farther, in South America. From this we infer that, in that period, similar geological influ- enees were exerted all along the extreme west por- tions of both North and South America.

8

X.

IT may not be out of place in this connection to notice briefly the coal deposits in the adjoining British possessions. On the Pacific coast are valuable mines of lignite on Vancouver's Island, which are extensive and worked. The coals mined at Nanimo on this island contain on an average 46 per cent of carbon, 32 of volatile matter, and 18 of ash. On Queen Charlotte's Island, 100 miles farther up the coast, lignite anthracite is found.

Coal of the Northwest.—Within that immense territory stretching to the Arctic Ocean from its southern base, that portion of the northern boundary of the United States that extends from Lake Superior to the Rocky Mountains, are very extensive fields of lignite. These are all located northward along the eastern slopes of the Rockies, and on the numerous head-streams of the Saskatchewan, on the sources of Lake Athabasca and those of the Mackenzie, and down the banks of the latter toward its mouth, and perhaps to the Arctic Ocean itself, as is the case north of Alaska. The beds of this lignite, as far as known, range from *two* to *eight* feet in thickness. Numbers of these have been on fire we know not how long, since it originated in spontaneous combustion. One was on fire in 1789, when Mackenzie explored the river, and was still burning nearly *forty* years afterward, as verified by Dr. Richardson, who visited the region. Large tracts of land are found where the coal has been burned out and left the clay soil of a red-brick color, and even melted the sandstone rocks.

Nova Scotia Coal-Field.—The coal of this region is bituminous, and is the product of the Carboniferous age; for this reason it is thought to be probably an extension of that of Pennsylvania, which lies in the same northeast-southwest direction, and the coal found in Massachusetts and Rhode Island may be the only links remaining of a lost chain. There are in this portion of the Dominion a number of localities, where coal is found, but not in available quantities, because of the thinness of the seams. The entire coal area—that is, where coal is found—is estimated at several thousand square miles, of which perhaps one thousand are available.

Albertite.—In New Brunswick, near the head of the Bay of Fundy, is a mine unique in its character, as from it is obtained a substance named (we presume from Prince Albert) *Albertite*. This material is found in a vein which fills a crevice between rocks; it appears to have been originally fluid, like petroleum, and under certain conditions was hardened, and seems to have been derived from the decomposition of vegetable or animal matter, or perhaps from a combination of both. The gas obtained from this substance has a superior illuminating power. Albertite is defined as "a material intermediate between the most bituminous coals and the asphalts," though it is free from mineral charcoal or earthy matter. There is a similar deposit in West Virginia (p. 60).

Pictou Mines.—On the north shore of Nova Scotia is the Pictou coal-region, within what is known as the curve of the coast. In this location are a number of seams or rather masses of coal, ranging in thickness from twelve to forty feet; and still another bed, 157 feet lower than the first, is about twenty-five feet in thickness; and a third seam, 280 feet below the second, has a depth of eleven feet. These facts have been ascertained by means of borings instituted for the purpose; but some authorities have doubted their perfect accuracy, as the beds have an inclination of 20°, while the measurements were in a line perpendicular from the surface. This coal is remarkable for the amount of ashes it contains, that being from 13 to 14 per cent, as well as sulphur to a small extent. Says Dr. Dawson: "The worst defect of Pictou coal is, that it contains a considerable quantity of light, bulky ashes, and this causes it to be much less esteemed for domestic use than on other grounds it deserves." The lower seams of this field extend apparently under the Atlantic Ocean and Cumberland Strait. The coal area of Pictou is said to be about twenty-eight square miles, but it would seem that the available space for mining is less than the estimate.

Sydney Mines.—The most important mining district in the Dominion, in respect to its output, has its center at Sydney, on the east shore of Cape Breton Island. The beds lie in basins along the coast for thirty-five miles, and from four to five feet in width. One seam or bed of coal is ten or eleven feet thick, and spreads back from the shore in a semicircular form; then comes a seam farther back, which is six feet in thickness; and still another, farther inland, eight feet in depth. These beds furnish a coal that is claimed to equal the famed Westmoreland, near Pittsburg (p. 53), in the production of gas. It is estimated that 120 square miles would be the extent of the workable seams in this coal area.

XI

THE COAL OF EUROPE COMPARED WITH THAT OF THE UNITED STATES.

IN two respects the coal-fields of the United States are strikingly in contrast with those of Europe: the one, in their greater surface extent, and the thickness of the seams; the other, in the mining of the coal with greater ease and at less expense. In Europe the seams of coal are more numerous, and as a general rule comparatively thin, and for that reason unworkable, and those thus available are the exception; while in the United States the seams are not so numerous, but the thin ones are the exception, and therefore a much greater proportion of the coal is workable. In Europe the coal-fields are comparatively small and scattered widely; in the United States, with very few exceptions, the separate coal-fields are immensely large, numbering many thousands of square miles.

Coal Areas.—The surface of Europe is estimated by geographers to be 3,750,000 square miles; that of the United States, by the same authority, at 3,000,000, excluding Alaska. The workable coal area of all of

Europe has been estimated to range from 15,000 to 20,000 square miles—we will take the latter number —and with an average depth of *ten 'feet* of available coal, not taking in the seams of different sizes that are too thin to be workable. From this we find that Europe, for one square mile of available coal, has 188 square miles of surface. The United States has, of the Carboniferous age alone, 205,000 square miles of coal area, estimated to be on an average *twenty 'feet thick.* (*Coal, Iron, etc., p. 320.*) In addition, the lignite variety, located west of the ninety-ninth meridian to the Pacific, is estimated at more than 150,000 square miles, but we will call it 100,000; that gives of the two classes of coal 300,000 square miles, which is at the rate of *one* square mile of coal to *ten* square miles of surface. The thickness of the beds of lignite is not so uniform as that of the coal of the Carboniferous age.

A **Special Comparison.**—As we are more intimately connected by means of commerce, and in having similar mechanical industries, with Belgium, France, and England, than with the other countries of Europe, we will briefly notice the coal found in the above-named countries. In Europe, seams of coal are worked that Americans at this day would deem unavailable, because of their thinness and the expense of removing the surroundings of the coal itself; but, labor being so cheap and coal so dear, in Europe mining under even such difficulties is made to pay.

Belgium is reported to have, in all, a workable coal area of 510 square miles. In its two most important coal-fields, the one in the vicinity of Liége has seams varying from six inches to five feet and a half—"the average being barely three feet"; the other, near Mons, has seams "from ten to twenty-eight inches—very few three feet in thickness." France has a coal area of 4,000 square miles—that is, where coal is found—but has only 1,000 square miles of workable coal. The country of Europe best supplied with coal is Great Britain, she having an area of about 12,000 square miles where coal is found : the portion of this area that is workable is estimated by one authority at 6,195, and by another at 5,000 square miles; the first estimate gives one square mile of coal to nineteen and a half square miles of surface, and the second twenty-four; Belgium has one in twenty-two and a half, and France one in 200. In England the thinness of the seams is so great that the miner has nearly always to lie on his side when working. As an instance of the labor in obtaining coal in some places, it is cited that "one seam of fourteen and another of sixteen and a half inches were worked to a depth of about 1,260 feet, but the circumstances were peculiarly favorable, . . . as under the coal was a soft clay that was easily removed." (*Coal, its History and Uses, p. 306.*) At Newcastle, in the richest coal-field of England, " the current thickness of the seams is from three to six feet "— though it is thought that " of British coal-measures, rather more than three fourths may become available for consumption." In Wales the seams do not exceed *three* feet in thickness ; and in one instance the coal has been mined a mile and a half under the sea.

In the anthracite regions of Pennsylvania, the main seam—" the Mammoth "—ranges from thirty to sixty feet, for the most part forty feet, without a break ; and the famous Pittsburg seam varies from eight feet to sixteen in thickness, with workable seams in many instances above and below it. A representative of this seam is found throughout the 205,000 square miles of the coal of the Carboniferous age. Seams of workable thickness predominate also in the lignite coal-fields of the Union. Another contrast is in our immense field of anthracite, of which none in comparison is found in Europe worth naming ; in addition we have large fields of semi-anthracite, which contains so much more fixed carbon than bituminous coals. We have five varieties —the anthracite, semi-anthracite, bituminous, lignite, and cannel.

Horizontal **Seams, and the Dip.**—With the exception found in the anthracite region, the coal-seams of the United States are uniformly horizontal or nearly so, and usually above the water-level, and in consequence they are easily drained, and the coal taken to the mouth of the mines in the side-hills on a level, and at comparatively small expense. When the coal, as in the central field, is under the prairie, it is reached by shafts which vary in depth from about

40 to 700 feet. The coal here lies also almost horizontally, and is easily mined, the galleries running in every direction from the bottom of the shaft, to which point the coal is brought and hoisted to the surface. The lignites in the foot-hills of the Rocky Mountains have more dip, but not sufficient to impede the mining by shafts to much extent.

In England, for instance, the dip or inclination of the coal-seams is so great, together with their usual thinness, that to work them is comparatively very difficult as well as expensive, especially in hoisting the coal to the surface and in freeing the mines from water, to do which requires the continuous action of the pumps. The coal lies very low and never horizontal; to reach it, shafts are sunk, and from these the miners follow the seams. Coal being originally formed on a level, the internal convulsion that tilted up the strata of the British Isles, and buried the coal so deep, must have been exceedingly violent— much more than in the anthracite region of Pennsylvania, where evidently was the greatest convulsion within the coal-fields of the United States. This may account for the reason that the coal runs so deep in England, where the depth ranges from a few hundred feet to the greatest that has been recently reached, that of 2,688 feet, more than half a mile.

How long will the Coal last?—There is much speculation in England in regard to the time when her coal will be exhausted. Sir William Armstrong is quoted as saying, in the year 1860: " Our [the English] total stores of coal, to the depth of 4,000 feet, will be exhausted in 212 years," at the present rate of mining. Two contingencies appear to conflict with this theory ; one of which is, will the coal run to that great depth? and the other, will the same thickness of the seams continue? To reach that depth the shafts would have to be sunk 1,312 feet beyond the deepest mine yet worked. In view of these considerations, it is " pointed out that actual exhaustion must take place at a much earlier period." Prof. Jevons says that " the coal of England, at the rate at which it is now [1863] mined, would be exhausted in less than 110 years."

In the anthracite region of Pennsylvania there are 472 square miles of coal area by actual survey. The amount of coal mined to the acre may be stated at 60,000 tons ; but there is an immense deal of wastage, so much so, that well-informed proprietors and scientific miners assert that careful and saving methods of mining could increase the product from the acre one-half—that is, 90,000 tons. The time may come when it will pay to be thus economical, and even to work over much of the territory. If, instead, we reckon the available product at 75,000 tons to the acre, an easy calculation will show that, on an annual average of 34,000,000 tons—about the amount mined in 1885—and making an allowance of fifty years for what has been already taken out, the coal of this anthracite region alone will last 616 years. (*Mineral Resources of U. S. for 1885, p. 1.*)

XII.

PETROLEUM, or rock-oil—a name derived from the Greek word *petra*, rock, and the Latin *oleum*, oil —is an inflammable oily liquid, that has been known to man from remote ages; in ancient history it was described as issuing from between the rocks. It evidently occurs in small caverns or reservoirs amid strata of rocks that lie in a horizontal position or nearly so, and apparently in long and irregular crevices. It is often found in connection with both gas and water—the latter being on the lowest level,

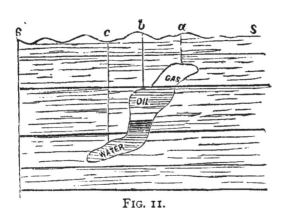

FIG. 11.

and the gas on the highest. If the auger in boring strikes the gas first, it rushes to the surface with terrific force. Sometimes the gas may be in a different crevice, but have communication through

a passage between the strata with an oil-reservoir;
if the latter is pierced, a flowing well is the result,
as the oil is forced to the surface by the pressure.
When that ceases, recourse is had to the pump.
Petroleum has evidently an intimate connection
with bitumen and bituminous coal, of whose vegetable origin there is no doubt, and yet there are
elements in petroleum that indicate additional ingredients to those belonging to bituminous coal.

Origin of Petroleum.—To account for this feature, the theory has been adduced that the elements
composing petroleum have been derived from animal as well as from vegetable matter, and "that bitumen and petroleum are formed from the more
perishable cellular plants and animals in the presence of salt-water." But there are objections to
this theory. Says Prof. Dana, of Yale : " The absence
of distinct fossil animal and vegetable remains points
to an abundance of delicate water-plants, or infuso-
rial or microscopic vegetable life, as the source of
this organic material." But oil, that in every respect is similar to petroleum, has been distilled
from both bituminous and cannel coals, and this
certainly is a proof of the connection. There were
in 1859 nearly sixty factories in the United States
engaged in distilling coal-oil, when in August of
that year an artesian well, to test the matter, was
sunk in the vicinity of an oil-spring near Titusville,
Pennsylvania. At the depth of about seventy feet,
oil was "struck," and soon it was pumped out at

the rate of twenty-five barrels a day; a year and a
half later a flowing well in the neighborhood sent
out three hundred barrels every twenty-four hours.
Here commenced the tremendous revolution in the
production of petroleum. The artificial product
ceased, and, instead, incorporated companies and
private parties sought the oil, which had been distilled, long centuries before, in Nature's laboratory,
and stored in vast reservoirs deep within the earth.
Then began the usual excitement and speculation,
and continuous over-production, accompanied by a
wastage that was sufficient to supply the wants of
the people for years and years.

Petroleum; **where found.** — In Northwestern
Pennsylvania and in Southwestern New York is an
oil territory extending into both States, the much
greater portion being in the former. This peculiar
oil exudes from the earth here and there in that
section, though usually in what are termed oil-
springs—the oil floating out on the water. Sometimes, instead of oil, gas would reach the surface
through springs, having been forced by the internal pressure along the route of the water. This
gas when lighted would burn, and these were called
burning-springs. In time it became known that this
gas was in some mysterious way allied with the oil.
In that section of country known as the oil-region,
oil is found; but there is no specific rule by which
to determine where it can be reached **by** boring,
and wells are sunk on presumptions, rather than

on scientific data: hence so many are failures, and, when "oil is struck," the reservoir is veiy uncertain as to the amount it may contain. These numerous districts where oil is found lie in a south-west-northeast direction for more than one hundred miles, but of uncertain breadth, and apparently lying in no regular order. Whole districts have sometimes become exhausted, while others are sought to take their places.

In the State of Ohio are several localities where petroleum is found, a prominent district being about twenty-five miles north of Zanesville. Also another location, in Hancock County, has numerous prosperous wells. Kentucky possesses oil territory to some extent, in which wells have been sunk and petroleum obtained. At present the most important district is on the border of the Tennessee line, known as the "Marshall tract," containing about thirty square miles of area. The amount of oil obtained here is comparatively small, and it remains to be ascertained if deeper borings will not produce more, as is often the case. Along the Big Sandy River are a number of oil-springs; the latter may lead to the discovery of a productive oil territory, while in the same State there are indications of a large district in Montgomery County. The southern end of the oil-belt extends southwesterly into Tennessee and across into Alabama, but has been as yet only partially developed.

West Virginia has an extensive oil territory, but

the production thus far is not very large, though the wells are deep. The production of oil has been pushed to such extremes that the glutted market will perhaps induce more discretion hereafter, and thus leave some for the use of succeeding generations; in this case excessive greed has wrought an injury that must be felt in the future.

Rocky Mountain Oil.—In this region petroleum is found in a number of localities. Colorado has extensive deposits in Fremont County; they are on Oil Creek, a few miles from Canyon City. Several small wells were sunk, and the oil produced found a market for its consumption in Denver. The production is still quite limited, but the field is by no means exhausted; some of the wells have been sunk from 1,700 to 1,800 feet deep. South of this deposit, some dozen miles, is another oil territory within the valley of the Arkansas River. The wells there have daily produced oil to a moderate amount. New Mexico has also indications of petroleum, but as yet it has not been developed.

Wyoming Oil.—This Territory has very valuable deposits of petroleum scattered through her wide domain. It is remarkable that here are required no deep borings, and yet the wells are very productive, while numerous "oil-springs" pour it out. The oil, exuding from the earth during the course of ages, has hardened into a crust partaking of the nature of bitumen, and has extended over much of this oil territory; but when the crust is probed, the petroleum shows itself. A singular specimen of this feature is in the Shoshone oil-basin, which has an area of about sixty acres, over all of which from time immemorial oil has been oozing, and gradually hardening into a crust from one and a half to three feet thick. But what is remarkable, over this crust has been deposited an alluvial soil several feet in thickness; and, wherever the crust is penetrated, the oil at once rises to the surface. Within this area a well was sunk, which, at the depth of 109 feet, and another at 350, reached apparently a large reservoir of oil, the latter pouring out eighty barrels daily. There are a number of isolated oil-basins, and of different sizes within the Territory, extending from east to west for 130 miles. In the northeast portion of the Territory, on the borders of Dakota, is also a large deposit of oil. No doubt there are many others yet to be found. Upon the whole, Wyoming has, as far as discovered, the most extensive oil-bearing area, of any State or Territory in the Union, while the oil itself is of the best quality. (*Min. Res. U. S. 83–1884, p. 217, and Report of the Governor, 1885.*)

California Oil.—The discovery of oil in Pennsylvania excited curiosity in the subject, and indications of oil were looked for all over the country. This led to the finding of oil in California (1859), which proved to be so abundant that in less than seven years there were seventy corporations engaged within the State in producing petroleum. In Hum-

9

boldt County alone were from forty to fifty wells; but there was not a flowing one of the number, and altogether they only pumped 1,200 barrels in one year. Speculation led to injudicious investments, and in less than three years the industry was practically brought to an end, but afterward renewed, new territory being discovered and utilized. The oil territory of the State is estimated at about 500 square miles, lying principally in the south-middle, and southern portion. The entire oil territory of the United States is estimated at 150,000 square miles.

Asphaltum.—This is a mineral pitch, or compact bitumen, which has an intimate connection with petroleum. This substance is put to many uses, such as for paving streets and sidewalks, for roofing buildings and floors, and for lining cisterns and pipes, etc. We have in conection with coal noticed the deposit in West Virginia, which thus far discovered is the only asphaltum found in a bituminous coal-field. In the lignite region of the Rocky Mountains several deposits occur in Colorado, and one exists in Idaho.

Southern California possesses immense beds of asphaltum ; but as a general rule it is not very pure, as when in a soft state it absorbs a large amount of extraneous matter. A deposit is near Carpenteria and immediately on the ocean-beach, and another covering a large area is located seven miles west of the town of Los Angeles ; while in Santa Barbara County the beds are not so large, yet they compensate in having an asphaltum of a better and purer quality.

XIII.

THIS gas has manifested its presence from time immemorial by issuing from fissures in rocks or in the crust of the earth, but more frequently, because less impeded, from springs, along whose course within the earth it has forced its way from the depths below. But how it is generated in Nature's laboratory, and stored in mysterious recesses within the earth, can be explained only partially by either the chemist or the geologist. It is very often found in connection with petroleum, and, when it is met with outside the locality of the latter, it is amid sandstone formations similar to those in which petroleum usually exists. At first when the latter was sought for by boring, and gas obtruded, it was deemed a nuisance ; but time and experience have changed all that. Gas-wells vary in depth, because gas lies deeper in some localities than in others, ranging from a few hundred feet to more than two thousand.

In 1878 parties were drilling for petroleum at Murraysville, eighteen miles from Pittsburg. " A depth of 1,320 feet had been reached when the drills were thrown high into the air, and the derricks broken to pieces and scattered around by a tremendous explosion of gas, which rushed with hoarse shriekings into the air, alarming the population for miles around. A light was applied, and immediately there leaped into life a fierce, dancing demon of fire, hissing and swirling about with the wind, and scorching the earth in a wide circle around it." (*Triumphant Democracy, p. 242.*) Thus, for five years, "the giant leaped and danced as madly as at first," when the gas was captured, and by means of pipes was conducted to the city and utilized in making iron.

Utilizing the Gas.—According to history, the first application in this country of natural gas—derived from a burning-spring—for illuminating and heating purposes, was made in 1821, at Fredonia, Chautauqua County, New York. This led to further investigation of the subject, but it took more than half a century to bring about its complete utilization, both for light and fuel. One of the first instances (1874) in which it was used in manufacturing, where great heat was required, was when it was brought in pipes eighteen miles to an iron-mill near Pittsburg; there it was "used under twelve boilers, in nine heating-furnaces, and in twenty-eight puddling-furnaces." This has within ten years from that time been far exceeded; one iron-establishment it is estimated consumes 1,000,000 cubic feet an hour—for meters are not used in ascertaining the amount of natural gas consumed. "When the Edgar Thomson works at Pittsburg are running to the full capacity, the gas consumed takes the place of 400 tons of coal a day." (*Min. Res. U. S., 1883–1884, p. 239.*) The utilization of this gas in Pittsburg and vicinity, since 1883, has been marvelous in extent, not only in manufacturing, but in the simplest form of domestic use. It seems, however, to be more effective in producing heat than in giving light, it having the former power more than the latter. It is very easily handled, and the heat generated being even sufficient for the purpose of making steel. Manufacturers of glass use it especially in making flint-glass, as the heat can be better regulated, while the burning-gas is free from smoke, which often discolors the glass. In making the finest fancy ware, it burns into the glass the beautiful flowers painted in different metallic colors; the process requiring, sometimes, as many as *five* separate burnings, according to the number of colors used.

Are the **Stores of Gas** exhaustible ?—Some of the wells in this gas area have been in use for a number of years, and the flow has not diminished perceptibly; while some have given out within a few weeks, others again have diminished slowly, and in the course of time have failed entirely. Often, when a well is flowing freely, if in its neighborhood another is drilled, the former shows at once a falling off in the flow; the inference is, that both are drawing from the same reservoir. When oil-wells cease to flow, recourse is had to the pumps; but there is no such remedy for the gas-well. Natural gas, in respect to its final exhaustion, is governed by laws similar to those that govern coal, as the latter does not itself increase, but confers benefits upon the current generation by being used up, thus fulfilling the object of its creation, and as there is no clear evidence that, in the laboratory of Nature, gas is still forming or making, it must in time, in the process of conferring these benefits, be also exhausted.

Where Natural Gas is found.—In respect to the area of territory in which natural gas may yet be discovered, a clew may be obtained in its intimate association with petroleum, and, where the latter exists, the former is likely to be found in the vicinity. We have no definite data to locate that area, and can learn of it only from time to time as discoveries are made in different portions of the country. When there is a call for its use, perhaps other areas will be discovered, that may be as valuable as the wonderful one in the vicinity of Pittsburg.

The production of natural gas is by no means limited to the vicinity of the Iron City, since it has been discovered in many places outside that region, and indeed from it far remote. Indications are that this natural resource, in due time, will be of still greater importance to the nation at large. In Northeastern Ohio, and in other portions of the State, gas-wells are in operation, and some of them have been for a number of years. In West Virginia, wells have been drilled and the gas used in lighting towns, and in some instances for manufacturing purposes. In the State of Kentucky, gas is known to exist in the vicinity of Louisville, and in other places, among which is Paducah. It is also found in the southern portions of the States of Indiana and Illinois. According to the newspapers of the day, drilling for gas has been successful in the vicinity of Chicago. The State of Alabama has natural gas, and so has the State of Kansas, and it is found in Davidson County, Dakota, in Wyoming Territory, and in Southern California. These wells may be the pioneer of discoveries equal in magnitude to those at Pittsburg and its neighborhood.

XIV.

WE come now to consider iron, of all the metals the best friend of man. It is the only one that can be *welded;* hence its immense utility in every form in which it can be manipulated, whether in making strong machinery, as the shafts of ocean-going steamers, or, when refined into steel, the most delicate surgical instruments, or the finest needle. Its great value to civilized man is so evident that political economists have sometimes taken its production and universal use as a criterion of the material progress of a community.

Theories as to the Origin of Iron.—Though iron is universally distributed throughout animate and inanimate creation, we shall not find its ores lying in the regular order as we found coal, but usually in beds or deposits, great and small, and also scattered and isolated. We can ascertain the ingredients and the formation of coal from the remains of the vegetable products of which it is composed; but we have no such clew to the origin of iron. " Iron-ores occur in so many different forms, and under so many chemical combinations, that no one theory of formation can cover the coincidents and conditions with which they are found." (*Coal, Iron, and Oil, p. 561.*) It may have been originally a simple and pure element, placed amid the other material of the earth, but by its affinity attracted to itself certain foreign substances, and was deposited under different circumstances in places almost innumerable. Though in theory it may be a simple element, and in minute particles, yet its combination with these various substances appears to have modified its character, and we find the iron and steel made from certain ores, tough and strong, while that from others is comparatively brittle and weak.

Names of Iron-Ores.—The reader may be interested in the characteristics and significant names of the various iron-ores that will come under his notice in this volume, and in the forms in which they are found in the mine. The most important of these ores, because of its good properties and being much diffused and quite abundant, is named *hematite*, from a Greek word meaning *blood-like*, as the ore is of a red color. Iron has a strong affinity for oxygen, and in hematite it is combined with three parts of that substance to two of itself. This ore is also called *specular*, when it is in crystals and has a bright metallic luster. Another form is called *brown hematite*, having the same proportions as the first, but in addition is allied with water, the common name of the latter being *bog-ore*, because found in bogs and swamps—the scientific one being *limonite*, derived from a word having a similar meaning; this ore often takes the form of black sand, and is usually magnetic. And, finally, *magnetic ore*, as it attracts iron and some iron-ores—the common name, lodestone. It belongs to the chemist to give the *rationale* of these various combinations, and to the geologist to designate the geological ages of their formations.

The localities in which iron-ores are found are very diversified. Some are in bogs and swamps; some in the mountains and hills, and often within the crevices of various kinds of rocks; some amid ledges of limestone or in lumps in the limestone soil; and some in connection with coal-measures.

New England Iron-Ores. — These States can boast but little of large deposits of iron-ore, and those they have are widely distributed, and isolated one from another, though for the most part they produce a remarkably fine quality of iron. The ores of that section belong generally to the limonite or bog-ore variety, where it is often found in the form of black sand, which is usually magnetic. Many of these beds of ore were discovered and utilized in colonial times and during the Revolutionary period, and immediately after. In numerous places are still to be seen the ruins of furnaces of the olden time, that have ceased work, either because of the exhaustion of the ore-beds, which was frequently the case, or the expense of obtaining charcoal from the forests on the neighboring hills or mountains. In 1734 iron was manufactured in Salisbury, Connecticut, and in 1784 there were in Massachusetts seventy-six iron-works, most of them being small. Certain ores in the western portion of the latter State produce so fine an iron as to hold its own in spite of the competition arising from districts west of the Hudson, in which is not only an abundance of ore, but also near by plenty of coal and limestone. For illustration, the ore deposit near Lanesboro furnishes an iron so tough that car-wheels are made almost directly from it. Upon the whole, the iron obtained from these isolated deposits of ore is remarkable for its fine qualities.

Connecticut has many of these peculiar deposits, especially in the town of Salisbury, in Litchfield County. This series of iron-mines extends into the neighboring counties of Columbia and Dutchess in New York State, and also into Berkshire in Massachusetts. The area in which these isolated deposits of iron are scattered is from ten to fifteen miles wide and perhaps from forty to fifty in length. The ores found in this somewhat noted region are brown hematite, but are different in their positions, some being in deposits and others in veins. The iron produced from the Salisbury ores was once in such high repute that the American navy was in a measure supplied with cannon that were made from it; but it is now principally used in the more humane and peaceful operation of making car-wheels.

The Great Iron-Ore Belt.—Before noticing in detail a few of the prominent deposits located in the famed Alleghany region, let us glance at the outlines of this immense belt of iron-ores. It commences in Northern New York, at the iron-mines on the west shore of Lake Champlain, and stretches toward the southwest, including the isolated deposits in the Cat-skill region, and in Northern New Jersey ; thence along the eastern slopes of the Alleghanies, includ-ing the numerous small valleys and hills running east and west, together with the famed Cumberland Valley and its extensions, the Shenandoah and the Tennessee, through Northwestern Georgia, taking in Northern Alabama, and then turns northeast along the west slope of the same range, across Eastern Kentucky, West Virginia, and Western Pennsylvania, including the iron-mines of Eastern Ohio. Between the eastern and western slopes within the mountains themselves are also deposits of iron-ore, coal, and limestone. The entire length of this mineral belt is more than a thousand miles, and the average width about seventy. What a contrast the Alleghanies present in the possession of this wealth of the most useful minerals, when compared with the Alps, the Andes, or our own Rocky Mountains and Sierra Ne-vada ! The iron and coal of this region alone, in their usefulness to man, are far in advance of the gold and silver of California and Nevada, of Mexico and South America. The application of the two former to the various kinds of industry will do more to advance the true welfare of the Nation than could be done by all the gold and silver of the world combined. " The annual output of the two precious metals, in the United States, is about the same in value as that of the pig-iron, but far below the value of the coal-production." (*Mineral Resources U. S., 1883-'84, p. 312.*)

New York Iron-Ores.—In the extreme northeast of the great belt are located the iron-mines belonging to the Lake Champlain district. These mines were worked in colonial times and produced " the best bar-iron which had ever made its appearance on the London market from America." (*Pearse's Iron-Manufacture, p. 64.*) These mines are on the west shore of the lake, and extend into the Adirondacks and occupy quite a large area. The ore is magnetic in its character and is rich and productive. There are numerous isolated iron-ore beds in the State, both east and west of the Hudson; one of the most interesting of these is at Stirling Mountain, in Orange County. This mine belonged originally to William Alexander, an officer in the army of Washington, but who also had a title, Lord Stirling—hence the name. This is a very large deposit of ore known as black oxide, which produces an iron equal to any in the world for its excellent properties, and in cousequence it has always been worked to advantage, as its yield averages 71 per cent of metallic iron. Mount Stirling rises about 400 feet above the plain, and on its eastern slope are the mines. It is noticeable that the rocks amid which this magnetic ore is found, geology says, are of the same character as those of the Danemora mine, in Sweden, which produces an iron famous the world over for its excellent qualities.

New Jersey Iron-Ores.—The northern portion of this State is full of iron-ore. Says an authority in speaking of this region: " The amount of ore developed within the ranges of hills in New Jersey is practically unlimited." It greatly enhances the value of the ore thus located that it is within easy reach of the anthracite coal-fields of Pennsylvania. The ores are easily mined, and the facility of thus obtaining fuel renders this mineral region one of the finest in the world for producing iron.

Pennsylvania Iron-Ores.—The magnetic ores just mentioned are in intimate connection with those in Pennsylvania, into which State they extend in a southwest direction, being mined in numerous places along that line, the beds varying in thickness from *two feet to fourteen.*

There are rich iron-mines in the vicinity of Reading ; and west of that city, near the town of Lebanon, is one of the most remarkable deposits of iron-ore within the State, inasmuch as it is somewhat allied with copper. This ore is different in structure from the other ores of that limestone region. The deposit is thought to be the result of an internal convulsion by which the three hills in which the ore is found were pushed up between or through the ledges of limestone. These hills are in height respectively 78, 98, and 312 feet ; in the latter is the principal bed of ore, which is 500 feet in its extreme thickness. Well may the miners estimate that there are 60,000,000 tons of ore "in sight"! This immense deposit is worked as an open quarry, and the ore is removed from the mine by relays of cars. The ore is magnetic, is in a single mass imbedded in a basin of trap-rock, and contains sulphur and oxide and carbonate of copper, but is virtually free from phosphorus.

From the Lehigh region southwestward, along the foot-hills of the mountains and intervening valleys, are numerous deposits of iron-ore extending across the State into Maryland; in the latter, magnetic ores are found, but in limited quantities. The magnetic ores from the Lehigh region, southward, degenerate somewhat in quality and quantity—the average percentage of metallic iron ranging from 37 to 66, when compared with their product farther north.

Virginia Iron-Ores.—In what is known as the Piedmont region, amid the head-streams of the James River, on the east slope of the Blue Ridge, are numerous deposits of iron-ore, that appear to run more than usual in regular line. These beds are in abundance, and the ore yields about 69 per cent of metallic iron, as found by analysis of a specimen from a mine near Orange Court-House. These ores are valuable for making steel, as they are virtually free from phosphorus. On the western slope of the Blue Ridge, in the Shenandoah Valley, are also extensive ore-beds of brown hematite of good quality; it yields from 56 to 60 per cent of metallic iron. These numerous iron deposits extend along the valley for one hundred miles or more. This valley is noted for its fertile limestone soil, and for its abounding ledges of the same stone. The iron-ores are on both sides of the valley, but those on the west side are not so rich as those on the east; though brown hematite, it yields only 46 per cent of metallic iron. These deposits of ores may be described as extending across the State in a northeast-southwest direction, and are known to be fairly rich in the production of metal, while they are also deemed practically inexhaustible, and contain every variety of ore for making iron and steel. The ores are easily mined, while at hand there is limestone to flux them if the fuel is brought thither. In the extreme southwestern portion of the State, in the vicinity of Wytheville, occurs an extensive deposit of red-hematite ore. "The vein is twenty-two feet in thickness, and yields 62 per cent of iron."

North Carolina Iron-Ores.—In this State, lying along the eastern foot of the Blue Ridge, are numerous deposits, consisting of black oxide of iron and of the magnetic variety, yielding 70 per cent of metallic iron, and of an excellent quality. The ore is often granular or friable, and resembles black sand, and produces an iron that partakes of the qualities of that of Stirling Mountain in New York State. These veins of ore are very extensive, and stretch across the State into Georgia. The whole region along the foot-hills of the mountain-range is peculiarly rich in minerals—copper as well as magnetic iron-ores and red oxides. These ores are now accessible to the coal-mines on the Kanawha, West Virginia, by the Ohio and Chesapeake Railway, along the cliffs of New River. This iron-ore field of the southern end of the Alleghany or Cumberland range of mountains would appear, from the class of ores and the immense area occupied, to be superior in that respect to any other portion of that extensive range.

In the northwestern portion of the State, South Carolina has magnetic and specular ores in inexhaustible quantities, on the western slope of King's Mountain. Brown hematite occurs also amid the mica slates of the same region, while bog-iron ore is found in nearly every county of the State.

East Tennessee Iron-Ores. — In passing over from North Carolina to the west side of the Blue Ridge we enter the East Tennessee Valley, there to find the counterpart of the iron-ore deposits we have just noted. These numerous and large beds of ore are of the brown hematite variety; they extend across the State in a southwesterly direction, along the western foot-hills of the Blue Ridge to Chattanooga, then across Northwestern Georgia and into Northern Alabama. In the vicinity of Chattanooga are vast beds of iron-ore; while in Georgia, on the head-streams of the Coosa River and along its west bank, these de-

10

posits can be traced for about 100 miles into Alabama.

Upon the whole, the State of Tennessee is remarkably rich in iron-ores, as it contains almost every variety—magnetic and hematite, limonite or bog-ore and specular. In addition to these is another valuable ore, the red fossil, a species of hematite, which is confined almost entirely to the northeastern portion of the State, but penetrates into North Carolina. It is remarkable that the latter class of ore is found near Lake Ontario in the State of New York; and again in Pennsylvania, and thence at intervals in a southwest line to North-Central Alabama. West of the Great Tennessee Valley are two others parallel with it; within the latter two are numerous beds of iron-ore: one valley, twenty-five miles long, runs northeast into Kentucky; the other, sixty miles in length, extends southwest into Alabama. The amount of iron here stored for future use is enormous. In connection with the coal-measures of the State, is also an abundance of the carbonate iron-ores. In the southeast part of the State "in some beds of ore, manganese prevails in such proportion as to make the manufacture of spiegeleisen or ferro-manganese a possible source of profit." The largest body of bog-ores or limonites in the State is found in Middle Tennessee—known as the Western iron-belt. This vast deposit covers irregularly an area forty miles wide, and extends entirely across the State from north to south—an area of about 4,000 square miles. (*Handbook of Tennessee.*) "In the manufacture of car-wheels and best refined bar-iron the ores of Tennessee have no superior." (*Mineral Resources of U. S., Survey, 1883-'84, p. 279.*)

Alabama Iron-Ores.—In this State is an immense iron-ore field known as Red Mountain ; this is an out-spur of the Alleghanies or Cumberland, and is thus named from the tinge given its soil by the red oxide of iron. It is a narrow ridge, twenty-five miles long and from four to five wide, and appears to have been pushed up in the midst of the Black Warrior coal-field already referred to (p. 62), and in consequence has coal on both sides. This deposit of ore is red hem-atite ; is very extensive, while the beds range from two to fifteen feet in thickness. Limestone is at hand, while the coal near by can be coked, and the ore produces a good soft iron from the furnaces.

Georgia Iron-Ores.—The main iron-ore district of this State is in its northwest portion, and in the shape of a triangle, the first side of which, running east and west, is the boundary-lines of the States of Tennes-see and North Carolina ; the second, being due south, is the Alabama ; and for the third side, the southeast, the prolongation of the Blue Ridge, trend-ing southwest. Within this triangle, which is drained by the Coosa River and its numerous head-streams, is a region rich in iron-ores. Here red or fossiliferous or dye-stone ore occurs in vast quanti-ties in beds, outcropping in sandstone ridges that en-circle coal-measures. Amid strata of shales and sandstones are from two to four beds of iron-ore; the thickness of these varies from a few inches to ten or twelve feet. These ores underlie an area of "not less than 350 square miles, including that portion of the region only where are outcroping beds, that are believed to be of workable thickness." (*Common-wealth of Georgia, p. 118.*) Here are found facilities for smelting—both coal and limestone; while the whole region is densely timbered, and the ores themselves are practically inexhaustible.

On the eastern part of this region, in the foot-hills of the mountains, are found deposits of specular ore; and in the lower portions, between the mountains, large but detached deposits of limonite or bog-ore. Magnetic iron-ores occur principally in two belts extending across the State : one of these follows in a southwesterly direction, the western base of the Blue Ridge ; the other extends all along the Chattahoochee Ridge. This ore is met with in fragments over a large extent of country, and there is good reason to believe that extensive deposits of it remain to be discovered.

Kentucky Iron-Ore Field.—In tracing northward the iron-ore deposits on the west side of the Alleghanies or Cumberland, we come first to Kentucky, which State has its southeast boundary along the crest of that range of mountains. In that section of the State, the rivers that specially belong to it have their head-fountains, such as the Licking, the Kentucky, and the Cumberland. On the head-streams of these rivers, and amid the foot-hills of the mountains, are numerous beds of iron-ore, so many that it is esti- mated that the area within the State where iron-ore is found is about 7,000 square miles. This ore was dis- covered and utilized by the early settlers, who had a furnace on Licking River toward the close of the last century. " An iron region, in length 125 miles by 50 in width, lies along the Cumberland River and its tributaries. Rich deposits exist over the whole district ; some of these ore-banks exhibit a thickness from 50 to 200 feet of ore overlying limestone." (*Iron Manufactures, p. 96.*) The ore is all brown hematite of different varieties, such as honeycomb and others.

It is evident that a series of iron-ore deposits in the Hanging Rock district once extended in the gen- eral direction of the Big Sandy ; as, nearly opposite its mouth, on the other side of the Ohio, is an extensive district containing the same class of ores, of which district *Ironton* is the center. Perhaps the interven- ing links of the chain of ore-banks were worn away by the waters of the Ohio River in cutting their way through. It may be added that "the iron district lying between Marietta and Portsmouth, Ohio, ex- hibits a development of the ores of the coal-measures equaled in no other part of our country."

West Virginia Iron-Ores.—We have already seen this State, in proportion to its size, represented as one of the richest in the Union in coal and other min- eral wealth. This statement is verified in the fact that within its boundaries are numerous and extensive deposits of iron-ores, which consist of all the varieties known. These stretch along amid the western foot-hills of the Alleghanies from the Kentucky line on the southwest to that of Maryland and Pennsylvania on the northeast. In connection with these iron-ores is an abundance of coal and limestone; the latter's influence upon the fertility of the soil is seen in the forests so remarkable for the large growth of the trees upon the hills, and in the abundant crops, especially in the valleys, when the land is properly cultivated. In giving an outline of the native riches of this State, an authority, Edmund Kirke, says: "It may be enough to state that the whole region between the Alleghany Mountains and the Ohio River is underlaid with vast deposits of all the more useful metals and minerals, richer and more accessible, than are to be found anywhere else on this continent."

Johnstown Iron-Ores.—Pursuing our course toward the northeast we reach Johnstown, Cambria County, Pennsylvania, on the west side of the Alleghanies in the valley between them and an outspur, the lower but parallel range known as Laurel Hill. Here is a striking instance in which the Creator has placed the ingredients near one another, that man may the more easily combine them for his benefit. For the mountain-sides in the vicinity are interspersed with veins of iron-ores, seams of bituminous coal, and withal ledges of limestone. Sixty feet above the highest coal is a double bed of compact carbonate iron-ore, an ore in connection with coal-measures and ledges of limestone—the total average thickness of the two combined being from two to three and a half feet, containing 51 per cent of metallic iron; in addition, brown hematite is mined in the vicinity. Owing to these facilities of materials at hand, the famous Cambria iron-works, located here, have turned out in one year, 81,006 tons of iron, and steel rails, besides an immense amount of other classes of iron. But these works lay under contribution other ore-fields, some of which are in the neighborhood, and some from the distant mines of Lake Superior; for science, by means of experiments, has demonstrated that often the best iron or steel is produced from the mixture of ores from different deposits.

Chestnut Ridge Ores.—On the west side of this comparatively low ridge, which runs parallel with the Alleghanies, we meet with a series of carbonate iron-ores. They lie in veins penetrating the sides of the mountain and also those of its foot-hills, and extend across Pennsylvania into West Virginia. Frequently there are several of these veins of ore lying one above another, and ranging in thickness from a few inches to two feet and a half. These ores produce only from 36 to 38 per cent of metallic iron. Of course, these deposits run parallel with the famed Connellsville coke-basin, which lies along the west base of the Chestnut Ridge, almost its entire length.

These ores were discovered and smelted, to be sure, in a small way, by the first settlers, as early as 1790; a number of furnaces were put in operation in that region within the space of twenty years or more. In the county of Fayette, it is claimed by some, were built and operated the first iron-furnaces west of the Alleghanies.

Ohio; Mahoning Valley Iron-Ores.—In Northwest Pennsylvania is quite a large area where iron-ore is found. These deposits are near the Ohio State line, and belong to the coal-measures or carbonate class of ores, as do the much more extensive and rich deposits belonging to that State in the valley of Mahoning River, a tributary of the Alleghany. In addition there is here also an abundance of coal, as already noticed (p. 65). In this valley iron-ore was discovered by the first settlers, and a smelting-furnace was erected in 1808, but the fuel used was charcoal. The State of Ohio has many large deposits and varieties of iron-ore; these beds are located in widely separated districts, and, in that respect, the State may be deemed rich.

Michigan Iron-Ores.—Hitherto we have found iron-ores in more or less proximity to the coal and limestone needed to smelt them. We come now to immense deposits near which is neither coal nor limestone, but the compensation is in the great facilities, partly by water, for transporting the ore to where it can be cheaply smelted. Hence these ores in enormous quantities are taken to the numerous furnaces on the Great Lakes, to Pittsburg and iron-works in Middle Pennsylvania, to Brazil, Indiana, and to many other important centers of iron-manufacture. By means of experiment science has shown that oftentimes the value of the iron produced is enhanced by the proper blending of ores found in different localities, which are often distant one from the other.

The iron-mines belonging to the United States on Lake Superior are in the Upper Peninsula of the State of Michigan; while on the northwest part of the same peninsula, and bordering on the lake, are also copper-mines, but southeast and south of these is an iron-ore region, altogether composed of about 6,000 square miles; that is where iron-ore is found in mines more or less rich. Marquette, on the lake, is the outlet for these mines, and has given its name to that ore region. These ores are of the usual varieties, but have properties, nevertheless, that belong to themselves alone. The soft hematites of this region are closely allied in character to the brown hematites of Eastern Pennsylvania; here are also found magnetic ores, red specular or specular hematite. The metallic iron contained in these Lake Superior first-class ores averages 63 per cent.

Amid one class of rocks—Huronian, according to geology—workable deposits of iron-ore are found in great abundance on the north border of the peninsula, and likewise in Canada on the north and east shore of Lake Superior itself. It would seem that the area now occupied by the lake was once a portion of a vast iron-ore field, and that the depression was made by some mighty convulsion of the earth. The area of the "Huronian" deposits is about 2,000 square miles.

Frequently in these mineral districts is found what the miners call "float-ore"—that is, fragments that through some force have become detached from the main ledge and removed, sometimes a mile or two, from the parent stock; they are known as iron bowlders. These fragments are of various sizes, often amounting to as much as a hundred tons. There are in this mineral district of the State of Michigan eighty-five corporations engaged in mining iron-ores, and in 1884 these companies shipped of these ores 2,455,924 gross tons (2,240 pounds to the ton), to be smelted in the furnaces of other States.

Wisconsin Iron-Ores.—The iron-ores of this lake region extend into this State, they having the same general characteristics. The deposits are extensive and are found in the interior of the State, principally in Dodge County, in the vicinity of "Iron Range." The formation is amid rocks, and the quality of the ore is good and the quantity abundant.

Minnesota Iron-Ores.—In the northeastern portion of this State, bordering on the northwest shore of Lake Superior, is an extensive region, comparatively sterile in soil, but rich in mineral wealth. It is known as the Vermilion Lake district and also as the " Iron Range," as the latter name implies; within it are enormous deposits of iron-ore. The length of this district or range is about seventy-five miles, while it varies in width from twenty to forty—the area being about 2,250 square miles. This region has been recently (1882) opened; but it has been thoroughly explored, and in nearly every portion iron-ore was found cropping out under different conditions, sometimes between rocks, and then again imbedded only a few feet below the surface. The prevailing ore is hard hematite, having characteristics similar to the ores found on the south shore of the lake, and, as the result of more than thirty assays of specimens taken from widely separated localities, these ores were found to produce metallic iron at a rate of per cent ranging from 65 to 70.

"Of the vast extent and uniform good quality of the iron-ores of this district there is no room for doubt. . . . All these ores are sufficiently low in phosphorus for Bessemer purposes, but they contain little or no manganese." (*Mineral Resources of U. S., 1883–'84, p. 266.*)

Missouri Iron-Ores.—This State is rich in coal, but richer in iron-ores, though they are unequally distributed over its territory, as in the northern portion there is very little iron, but an extensive field of coal. In the south central half of the State, below the Missouri River, are valuable deposits of bog-ore. The latter are found in the region between the Osage River and the Ozark Mountains, as well as red hematite on the upper streams of the river.

There are three iron districts in Missouri: the eastern, including the Iron Mountain specular ores; the central specular ores; and the western or Osage region, with its immense deposits of bog-ore and red hematite. These districts form a broad ore-belt extending across the State from the Mississippi to the Osage, and south of the Missouri, but nearly parallel with the latter river. The bog-ores lie both northwest and southeast of the central or specular-ore district. Within this region have been discovered thus far nearly 300 deposits, great and small, of iron-ores of the three different kinds mentioned above. It is announced by geologists that all undisturbed bog-ore deposits in the State are resting on limestone, and the latter often even of different geological formations. This ore lies in irregular cracks, pockets, or crevices, either on or near the surface of the various limestones. All along the Mississippi in Southeastern Missouri are found these deposits of bog-ore, in beds ranging in thickness from a few inches to four feet.

Iron Mountain.—The most remarkable feature of the ores of Missouri is that of the famous Iron Mountain and Pilot Knob deposits. These vast treasures lie in a southwest direction about seventy-five miles from the city of St. Louis. The former is dome-shaped and in height about 250 feet above the plain, and is entirely covered by what is termed " surface ore " in the form sometimes of " moss-grown blocks, some of which are many tons in weight." These uncovered, loose, and small pieces and blocks of ore have been thus exposed by the wearing away in the course of many centuries of the earthy matter in which they were once enveloped. The internal structure, as far as ascertained by mining operations, shows the mountain to be composed of iron-ore in connection with porphyry. On the summit, as has been noted, is an enormous mass of solid ore, and veins of the same kind, but in different sizes, run in all directions. Iron Mountain, and its immediate surroundings of lower hills, constitute the largest deposit in the State. The ore is very uniform in its characteristics, though in the veins and beds the forms are not regular. The ore is magnetic, and frequently has even a distinct polarity like the needle, and yields about 70 per cent of metallic iron.

Pilot Knob.—Six miles distant from Iron Mountain is a hill shaped like a cone and nearly circular, known as Pilot Knob. Its diameter at the base is about one mile, and its height above the plain is 662 feet and 1,521 above the ocean. The Knob has several beds of hard specular ore. There is one of these forty feet in thickness, but separated in two portions by a slate-seam of ten inches up to three feet. The lower portion is thirty-one feet in depth and "is very compact, dense, and the ore hard"; it contains 60 per cent of metallic iron, while the upper one yields only 53 per cent, but each has very little sulphur or phosphorus, their principal impurity being silica. The ore of Pilot Knob is specular, but differs from that of Iron Mountain, and some others in the State, in being for the most part unmagnetic. It is estimated that nearly one tenth of the bulk of these two mountains is pure ore.

Texas Iron-Ores.—Deposits of iron-ore are found in a few places within the State, though they have not been much developed. In Rusk County, it is said, there are extensive mines of iron-ore of excellent quality. In the northern part of the State within the counties of Jack and Young, and between the rivers Brazos and Trinity, is a large deposit of iron-ore. The ore has been pronounced of excellent quality by experts. It has not yet been developed. The latter have been utilized, perhaps, more than any others in the State.

Arkansas Iron-Ores.—In Independence County, of this State, a very valuable discovery has recently been made of an iron-ore—manganese—that is suitable for the manufacture of spiegeleisen. This deposit is so immense in quantity that, of itself, it could render the United States, in the manufacture of that species of iron, independent of the rest of the world.

New Mexico Iron-Ores.—The deposits of iron-ore in this Territory are numerous; they extend from the

Raton Mountains, on the borders of Colorado, across the region drained by the head-streams of the South Canadian River, a branch of the Arkansas, and farther south through the middle portion of the Territory, on both sides of the Rio Grande in the vicinity of Santa Fé, and down south beyond Socorro. The ores located near Santa Fé are magnetic, and are also in the neighborhood of both coal and limestone. This may likewise be said of the ores that are found in Bernalillo County, which possesses fine seams of coal. It is said that all the iron-ore deposits of the Territory are in the vicinity of an abundance of both coal and limestone. With these advantages, this Territory at no distant day may take its place as a successful miner and manufacturer of iron.

East of the Gallinas Mountains, and about thirty miles west of the town of Socorro, is an extensive district covered with *nodules* or rounded masses of iron-ore, weighing each from 100 to 500 pounds, the ore itself being quite pure and productive of metallic iron. Excellent iron-ore is found, also, in several mountain-ranges in the southern portion of the Territory.

" It is stated that iron-ore occurs in many places in Arizona, and some very fine beds of hematite are reported in Gila County. But as yet none of the beds are worked." (*Mineral Resources of U. S., p. 289.*)

XV.

Colorado Iron-Ores.—There are many deposits of iron-ore in this State ; of these a few have been opened in the valley of the Arkansas. The Hot Springs mines, of which there are a number, are located on the western slope of the Sangre de Cristo range. The ore is a porous brown hematite of pure character, occurring in separate masses on the slopes of the foot-hills, there being no regular or continuous veins.

There are, also, quite extensive beds of iron-ore south of Leadville, near Gunnison City, in the valley of the Gunnison, a branch of the Colorado. The ore found here is rich, producing 68 per cent of metallic iron. The amount of iron-ores is so large in the region south of Pike's Peak as to warrant the establishment at South Pueblo of very extensive works for the purpose of manufacturing iron and steel, and also for rolling plate-iron and making nails.

Utah Iron-Ores. — This Territory has beds of iron-ore in numerous places within its boundaries.

Some of these ores, having been analyzed, were found to contain from 50 to 65 per cent of metallic iron. One extensive deposit is about sixteen miles long by three wide. Amid this area stands Iron Mountain, which rises 1,500 feet above the plain. The ore is hematite and magnetite ; the latter attracts the magnet, but does not itself become a magnet, having two poles. Some of this ore furnishes metallic iron of from 61 to 64 per cent. Iron deposits appear to be well distributed in Utah amid the foot-hills of the ranges of mountains, and are worked to a moderate extent compared with some of those in other Territories.

Wyoming Iron-Ores.—There are three kinds of ore in this Territory : the magnetic, the hematite, and the carbonate—the latter two being the more prominent. An important bed of hematite ore is at the base or foot-hills of the Seminoe Mountain, a few miles west of the Platte River, and about twenty-five miles north of the Union Pacific Railway. The ore, as proved by analysis, is of a superior quality, while a " scientific measurement of the area it ocenpies shows it to be practically inexhaustible." The carbonate ores of the Territory are superior and valuable, as they contain the elements so important in the manufacture of iron and steel by the Bessemer process. In addition, in their vicinity is a vast field of good lignite coal " for fuel and coking and plenty of limestone for fluxing purposes." The Territorial geologist—Prof. Aughey—as quoted by the Gov-

11

ernor in his report to the President, says: " An abundance of the best of ore, easily accessible, and fuel and fluxes, also close at hand, are present at the southwestern base of the Seminoe Mountain to a degree rarely found elsewhere." In Laramie County, about forty miles northwest of Cheyenne, occurs an enormous mass of iron-ore, so extensive as to give the name Iron Mountain to the district. " This locality is capable of furnishing indefinite quantities of iron-ore." (*Mineral Resources of U. S., 1883-'84, p. 285.*)

California Iron-Ores.—Rich deposits of hematite, magnetic, and other ores, are found in this State, and indeed this mineral occurs in more or less quantities in twenty-one of the fifty-two counties in the State. An extensive mine on the American River, in El Dorado County, has a vein about three feet thick, the ore of which carries a high percentage of metallic iron. Along many of the beaches of the coast are also found magnetic-iron sands ; these are extensive, though they have not hitherto been made very available, whatever they may become in the future. The people of the State have been in the past so much engaged in the more profitable business of gold and silver mining that, with the exception of a few notable instances, they have postponed the development of their wealth in iron-ore. Thus far California has depended rather upon her neighbors for this metal than upon her own furnaces, meanwhile importing the greater portion of what she used.

Oregon Iron-Ores.—This State has in masses the

usual iron-ores, bog or limonite, hematite and magnetic, all of which can be easily mined. Beds of ore extend from a point opposite Kalima, on the Columbia River southward, and also on the Willamette almost to the falls. To smelt the latter ores, furnaces have been established on the river a short distance south of Portland. Iron-ores in numerous and large deposits are located in several of the eastern counties of the State, all waiting to be developed.

Washington Territory Iron-Ores.—The Puget Sound basin has, throughout its entire extent, numbers of immense beds of bog-iron ore of the best quality, especially in the counties of Jefferson, King, and Pierce. This bog-ore is mixed to advantage in smelting with hematite, " brought from Texada Island, British Columbia, and limestone from San Juan Island, in the American group, the latter used as a flux, thus producing an excellent article." Brown hematite of fine quality has been discovered in Skagit County, and magnetic ore, also, in King County. In the same region "the Derry Mines extend almost due north and south, and are nearly vertical. The veins range in thickness from six to 150 feet. The thickest vein is magnetic iron-ore, the richness of which in metallic iron, and its almost absolute freedom from all deleterious substances, render it extremely valuable, especially for the manufacture of Bessemer steel." Six specimens of this class of ore yielded from 67 to 70 per cent of metallic iron.

XVI.

IT will be noted that in the United States the deposits of iron-ore are very much diffused, and sometimes remote from one another. These ores differ somewhat in their characteristics. Some, when unmixed with others of better quality, do not produce a specially good iron, and some are suitable for the manufacture of the higher grades of iron and of steel ; the latter class appears to be confined to limited areas, though they abound in quantities sufficient for the Nation's use.

In nearly every instance, within our wide domain, we have seen that, where iron-ore exists, in the vicinity is found plenty of fuel—it may be coal, or wood to make charcoal—and also limestone, so essential for the fluxing or causing the iron melted in the furnace to flow easily into molds as the intense heat compels it to exude from the baser material infolding the primitive ore. One remarkable instance of the absence of both coal and limestone occurs in the Lake Superior iron-region ; but that is quite compensated by the cheap and unrivaled facilities for transporting these ores to where the former two can be easily obtained.

Spiegeleisen.—Within recent years discoveries of beds of a peculiar iron-ore, and experiments on the same, have demonstrated that the United States have in abundance ores suitable for manufacturing Bessemer steel. Experts declare that " we can make as good iron for crucible steel as the Swedes do, if we would only be as painstaking as they are." In addition, we can supply ourselves from our own mines in the production of *spiegeleisen* and *ferro-manganese*, and thus be in that respect independent of the outside world. We now export manganese to Great Britain to be used in making steel, while our own manufacturers of that article use domestic ores.

The Change in the Iron.—Within recent years a revolution has taken place in the means employed to extract the iron from the ore; but still more strik-ing have been the modes introduced by which the iron itself is refined, until it takes the form of steel: and thus for all practical purposes the value of the metal has been greatly enhanced, so that now Bessemer steel is largely used for purposes in which formerly rolled iron was employed ex-elusively. Once it took weeks, even after, with many manipulations, the iron had been prepared for being refined into steel; but to-day, by the Bessemer process, at the second or third remove from the ore, that result is attained and within a few hours, while under certain circumstances, at the third or fourth remove, the iron is made into steel rails. Because of its strength being so much greater than of the old-fashioned iron, and of its less weight and greater durability, steel is used, instead of iron, for almost every purpose, and in the end is cheaper to the consumer. The being made of steel adds immensely to the usefulness of the ordinary utensils of the household of the mechanic and of the farmer. The latter derives very great advantages in the various combinations of steel employed in the manufacture of his farming utensils; witness the simple but light and strong steel hoe and the steel plow, compared with the cumbersome ones of the olden time, not to speak of its use in the machines which make the farmer's labors of to-day light compared with those of former times: even from the toil of preparing the ground for the seed and in harvesting the crop, and making it ready for the mill or for the market.

We have already noticed (p. 103) our immense resources in respect to coal when compared with those of Europe, and in relation to deposits of iron-ore we appear to be equally as much in advance.

The Bessemer Process.—In relation to the mode of making steel, the reader has noticed the use of the term " Bessemer process." That term denotes in a remarkable manner the application of science in refining iron into steel, and thereby greatly enhancing the practical value of the iron itself. The name is derived from that of the inventor, Sir Henry

Bessemer, an Englishman, though of German descent. "The old-fashioned way of manufacturing steel was first to produce wrought-iron in bars, to be placed in a furnace and imbedded between layers of charcoal, and then subjected to the heat of the burning charcoal, the air being excluded; this process continued for at least two weeks, and often longer." From this steel thus prepared was made *cast-steel.*

After experimenting for about ten years, Sir Henry at length discovered a process by which iron is refined into steel usually at the second remove from the ore: accomplishing in about *thirty minutes* that which formerly took two or three weeks! The new method removes the impurities in the iron by literally burning them up and foreing their *débris* out of the metal. This end is attained by the application of intense heat to the raw or pig iron, by means of a peculiarly constructed vessel called a "converter"—as it *converts* iron into steel. This huge (almost pear-shaped) crucible is swung on hinges or trunnions, and is moved by machinery, so that it can be turned over on the side to receive the red-hot pig-iron, and then placed upright, after which it can be so turned as to pour out the molten metal into the ladle. The converter is peculiarly constructed: in the bottom are a hundred or more holes in which are inserted tubes about one fourth of an inch in diameter, through which air is forced at a pressure of twenty-five pounds or more to the square inch. To be prepared for the reception of the raw red-hot iron the converter itself is heated to a white heat; it is turned on the side and the iron put in, and, after the draught of air is applied, it is turned into an upright position—the tremendous blast preventing the obstruction of the air-tubes by the melted metal. Now is exhibited, on a small scale, a peculiar volcano. The roaring noise made by the air-blast, and bubbling of the molten mass, can be heard for a long distance, and a stream of sparks made by the consuming impurities, and issuing from the mouth of the converter, blinds the spectator with their dazzling whiteness. In less than thirty minutes the combustion is completed, as indicated by the sparks ceasing, but followed by a lurid translucent glare of light from the molten metal now in a glowing white heat. The iron is now nearly pure and in a fluid state, so that it can be poured almost as easily as water. There remains still a small quantity of the oxide of iron, to remove which a sufficient amount of spiegel-iron is run into the mass. The manganese of the latter at once decomposes the oxide, and, taking up the oxygen, frees the iron, while it itself passes into a slag or oxide of manganese, which is skimmed off. A portion of the manganese remains in the mass, "giving it its steel properties."

The combustion being complete and the metal ready for the molds, the converter is turned and its contents—now molten steel—are poured into an immense ladle, which is also moved by machinery, and which in turn pours the liquid metal into the appropriate molds. These ingots of steel are at once ready to be transferred to rollers and made into rails, or to the great hammers to be fitted for other purposes. The American process " is a total renovation of the English method, and a great improvement." (*Knight's Mechanical Dictionary.*)

Manganese.—As we have just seen the great value of manganese in refining iron or converting it into steel, the reader may not deem it out of place if we briefly notice this (comparatively speaking) recently discovered metal. It is found in combination with several other minerals; traces of it exist in the soil and also in plants. In a pure state its color is a " dusky, whitish gray; it is very hard and difficult to fuse"; in this form it is of little or no utility; but in certain combinations is very valuable, as we have seen, in the manufacture of steel. It is also useful in several other combinations, such as in the making of glass, when it neutralizes a green tint derived from the presence of iron, or when under certain conditions it aids in obtaining chlorine and bromine, and in producing certain pigments; but all these are as nothing compared with its immense utility in making steel. The mystery involved in its thus refining iron belongs to the metallurgist or chemist to explain.

As a general rule, where iron-ores occur there is a

small percentage of manganese; but its peculiar usefulness remained unknown, and its special virtue in
refining iron was unsuspected. It had been noticed
that certain ores made better iron than others, but
why, was a mystery. At length it was discovered
that such ores had in combination with them more
than usual of manganese. This elicited the surmise
that the latter metal might have something to do
with this refining process; if so, why not supply the
manganese, where there was a deficiency, and test
the matter? The culmination of these experiments
is in the Bessemer process, in which, after the impurities incident to ordinary iron-ores are removed,
the manganese, in correct proportions, is put into the
" converter," and steel is the result.

Discovery of Manganese-Ores.—Thus was demonstrated the great importance of manganese in this
special industry. The American manufacturers of
steel soon took measures in order to ascertain if ores
producing that metal could not be found within the
United States. They wished to relieve themselves
of the immense financial burden in using vast quantities of imported manganese, should they engage extensively in making steel, while perhaps they had
within their own land the ores from which they
could obtain an abundance. The result of this
effort has been, that deposits of the required ore
have been discovered within the Union, and that
they are unusually large when compared with those
of Europe. Some of these ores had been noticed by

those who were ignorant of their qualities which science had not yet revealed. The manganese deposits
are irregularly distributed along the Atlantic slope.
In Maine are a number of mines of bog-manganese
ore; such deposits are also found, but sparingly, in
all the New England States except Connecticut, and
in limited numbers in New York and Pennsylvania.
Manganese-ores occur in New Jersey, but in combination with zinc in that singular metal known as
Franklinite—thus named from the Franklin furnace,
in the vicinity of which it exists. From Maryland
to Georgia, the ore called black oxide of manganese
prevails more than the bog variety; the former is
mined to much greater extent than the latter. In
the Shenandoah Valley, in Augusta County, Virginia,
is a valuable mine—the Crimora—from which at
first the ores were shipped to England and Belgium,
but recently they have been utilized in making spiegel or specular iron. This mine has an ore that contains nearly 89 per cent of manganese. Some of the
deposits in the State are quite large; but to none of
them could the term inexhaustible be justly applied.
Manganese-ores also occur in both the Carolinas,
though, thus far discovered, they are of a moderate
grade, and in the " Etowah region," Northwest Georgia; the latter ores are utilized in making spiegel-
iron, their average per cent being from 66 to 80.
Similar ores occur in West Virginia, in the vicinity
of Harper's Ferry. Many deposits are found in Alabama; the principal one yet discovered is in Calhoun

County ; manganese-ores have also been discovered in Dickson County, Middle Tennessee, and also accidentally found in Independence County, Arkansas. The latter is of the black oxide variety, and occurs in " pockets " ; but the belt of territory in which these isolated pockets are found extends for some fifteen miles, while the width is from six to eight, having an area, perhaps, of 100 square miles. Much of this ore is on the surface, or slightly imbedded, as if its original surroundings had been washed away in the course of ages. The cost of mining is, therefore, comparatively cheap, while the average percentage of metallic manganese is 54. This, taken altogether, may be deemed both a rich and extensive field.

Manganese on the Pacific Slope.—Large deposits occur in California, in Nevada, and more or less in the Rocky Mountains. In San Francisco Bay is a deposit of ore on a small island some ten miles north of the city, which has been opened to some extent. The island contains twenty-seven acres, and rises 250 feet above the water. The ore occurs in large masses, a heavy belt of which extends for seven or eight hundred feet across the island. In Tuolumne County, pieces of ore weighing 100 pounds have been found on the surface of the ground, the deposit being similar to that in Arkansas. There are quite a number of localities in the State where manganese-ores are found, but they are widely separated ; and in time other deposits may be discovered in the intervening spaces. The main deposit of ore in Nevada is in

Nye County, and is deemed valuable, but hitherto has not been utilized to any extent.

Spiegel-Iron **and** Ferro-Manganese.—The distinctive feature of these two composite metals is the combination of iron in certain proportions with manganese. When the latter is less than 20 per cent, the metal is called *spiegel* iron, from the peculiar specular or crystalline feature imparted to it by the manganese, and when the percentage is greater than 20 it is known as *ferro* (iron) manganese—"a metallic substance as essential for the manufacture of mild steel as spiegel is for steel rails." The latter combination, according to the mode of manufacture, contains intermediate grades of percentage from 20 up to 90. Manganese, owing to the many discoveries and experiments of chemists and metallurgists, has become very valuable because of its susceptibility of making alloys with so many metals, such as iron, copper, zinc, etc. The most important by far of these alloys is that with iron, as shown by its imparting to it, in the form of steel, numerous characteristic properties in proportion to the percentage in which it is introduced, making the steel thus modified exceedingly valuable for many purposes for which ordinary steel is not used. But we can not go into details.

It was found almost impossible to apply pure manganese in the manufacture of steel, because of the difficulty of melting it in order to secure its assimilation with the molten iron. Recourse was had to spiegel-iron, as it is rich in manganese, having been made by fusing together the oxides of the ores of iron and of that metal, and in addition experiment had proved that the manganese in this combination would, for some unexplained cause, assimilate with the molten iron in the converter, and impart to it an element that changed it into steel. It was of vast importance to American manufacturers of steel that they should be able to supply themselves with manganese from their own resources. Hence energetic efforts were made to discover, in their own country, ores producing spiegel-iron ; and we have seen with what remarkable success these efforts have been crowned. Until they made their own, they imported spiegel-iron from Europe—the largest quantity in one year being 25,000 tons. In 1870 the making of spiegel-iron was commenced at Newark, New Jersey, and the article produced was deemed equal to the imported. Numerous American establishments are to-day engaged in the same enterprise, and supply themselves with ores from their own ample mines. England now imports from us manganese-ores.

Duties due to Future Generations.—It is not presumed that, because ores of a metal are within reach, for that reason they must be mined. The people of the future have an interest in this matter, and that interest or right ought to be respected. That parent would justly merit the condemnation of the intelligent and the humane who, in using his property, should willingly waste it to the detriment of his children. Under such circumstances he would be looked upon as a monster. May not there be quite a parallel case in one generation wasting the heritage that jointly belongs to future generations? If this wise and humane principle were carried out in practice, would we see our forests dwindling away in wanton destruction; the wastage in our coal-mining operations; and the selfish capturing of fish and cutting off the sources of supply for those who are to come after us? In what category of patriots shall we place those legislative bodies and officials who, in control of the Nation's affairs, should have an eye only to the selfish advancement of the present generation, but at the expense of those that are to follow —that would neglect, for illustration, to use without wasting our numerous natural resources, or to put in train influences of education that would mold the present rising generation, enabling them as parents to acquire tastes that would insure their making appropriate efforts, in order that their own children might receive better intellectual and moral training than they themselves enjoyed? Such influences would necessarily go on forever, from generation to generation.

XVII.

GOLD.

How evenly Nature has balanced her gifts of metals to the American people! The eastern and the middle portions of the Union are lavish with their wealth of iron, while the western is equally as bounteous in its gold and silver. The comparison may extend still further, and as the former furnishes but little gold and no silver, so the latter yields but a small amount of iron.

Characteristics of Gold.—Gold is always found pure in nature, and in its native state has no affinity for other substances. Thus when imbedded in a kind of quartz known as gold-bearing, and when the latter, by the action of the elements, becomes disintegrated, the gold is left free, and is often found in small particles in the beds of streams or where the gold-bearing rock has been dissolved. Often, in tracing these streams to their fountains in the mountains, the latter are found to have among their strata gold-bearing quartz. Though having in its native state no affinity for other substances, yet science enables us, in preparing gold for the mint, to combine it with copper in such proportions as to render it sufficiently hard to diminish much of the wastage incident to its being used as coin.

Gold on the Atlantic Slope.—There are two gold-bearing areas in the United States: one—much the smaller—on the Atlantic slope, the other on the Pacific. Gold has been found in the State of Vermont, but in very limited quantities, and as yet attempts have failed to find it in any appreciable amount in the original quartz-rock whence it came. In one instance, among others in which mere particles were picked up, was found a lump of pure gold weighing, it is said, eight ounces and a half, to which were attached a few small quartz-crystals. "The gold formation of Vermont is a narrow and irregular belt extending through the entire length of the State." (*Whitney's Metallic Wealth, etc., p. 124.*) But the precious metal is not obtained from the original quartz-rock in paying quantities.

Particles of gold have been picked up in the State of Maryland, and also in Virginia; the latter finding is mentioned by Thomas Jefferson, in his "Notes on Virginia." In these he states that a lump of gold, weighing seventeen pennyweights, was found near the Rappahannock River; but that may have been lost by an Indian, and who perhaps obtained it from North Carolina, as, from what the Indians told De Soto concerning the refining of gold, they appear to have known of it in that region. (*American People, p. 56.*)

All along the Atlantic slope, from Vermont to Georgia, in the places where traces of gold are found, "the geological structure is the same from one extreme to the other," while "the matrix"—the substance in which metallic ores are found—"of the gold is invariably quartz." Half a century ago two gold-mines were worked in the Piedmont region of Virginia, and these in twenty years increased to a dozen or more; but none of them were in a true sense successful, because of the lack of the precious metal itself: in consequence, these mines have long since been virtually abandoned.

North Carolina Gold.—The region most productive of gold on the Atlantic slope is in North Carolina, though it is also found in South Carolina and Georgia. In 1799 quite a large lump of gold was picked up in the State first mentioned, and this discovery was the first indication given of its presence in that region. Mines were discovered in the course of years and worked in a crude sort of way, but for a quarter of a century the gold was obtained principally from washings. At length it was ascertained that the precious metal was in the veins of certain quartz-rocks, and attention was directed to these "vein-mines." In time there were nine different localities in which gold was found, and where mining operations on a comparatively small scale were carried on, sometimes at great expense and not with corresponding profit. These mines were worked till the discovery of the much greater producing gold-fields of California, at which time they were nearly abandoned. Since the recent war more effective and scientific methods have been introduced, and the mines, especially in North Carolina, have again been worked to better advantage. "The gold of this State is generally more or less alloyed with silver, varying from pure gold on the one side to pure silver on the other."

Says Mr. James P. Pomeroy, of Greensboro, a scientific miner: "In what is known as the Piedmont region is a belt of quartz-veins much disturbed by the rolling of the ground, so that there is no regular dip to the veins," which are often interrupted with breaks or faults. This belt extends from north of Greensboro in a southwest direction along the eastern foot-hills of the Blue Ridge to Charlotte. "Some of the ores are of good grade, as high as one hundred dollars to the ton, some even more, but much of it ten to fifteen dollars to the ton." South of the belt mentioned is another, which is not nearly so rich; and still farther south another, characterized by *talc-slates*, which carry gold rocks that "will yield by simple stamping from three to five dollars a ton. These ores are generally termed *refractory*, from the nature of the sulphurets with which they are combined."

South Carolina and Georgia Gold.—South Carolina has a limited gold-bearing area, and its mines were nearly abandoned, when a discovery of a rich deposit in 1852 suddenly placed her in the front rank of the gold-producing States on the Atlantic slope. This was the famous Dora mine, thus named from the owner. This mine in nearly a year and a half produced $300,000, the expense of operating being about $1,500. The gold was obtained in an excavation, three hundred feet long, by twelve deep and fifteen wide. Nuggets were sometimes found that weighed as much as sixty pennyweights, or five pounds (troy), but the deposit was speedily exhausted. In Georgia the yield of gold was at one time half a million dollars a year, but its mines also soon began to give out. Small quantities of gold have been produced in the States of Alabama and Tennessee, though the yield has grown less and less.

The amount of gold obtained in that region was so great that, in 1838, the Government established two mints, one at Charlotte, North Carolina, and one at Dahlonega, Georgia; these have long since been discontinued, the mines having virtually become exhausted.

California Gold—the Migration.—The discovery of gold in California, in 1848, virtually revolutionized the trade and the industries of the United States. Within less than two years, tens of thousands migrated thither from the older States. They went by different routes. Some passed in sailing-vessels around Cape Horn, and some crossed the Isthmus of Panama and found their way up the coast, while others went by a toilsome journey across the plains, aided only by the most primitive appliances for travel. All these left vacancies in the various departments of industry at home, which were filled by others at advanced wages. The knowledge of the discovery, even before the effects were fully realized, stimulated the main industries of the country in anticipation of the grand results, which in the end were beyond what was reasonably expected. These effects were produced by means of the immense amount of gold thus brought into circulation. That being the standard of value, prices of every kind seemed to rise; but the solution was rather in the fact that gold, because of its abundance, had become cheaper.

Inroads of Foreigners. — The news spread throughout the world, and " then commenced a wandering of nations," and chiefly foreigners were the first to reach the land of promise, coming, especially, from Mexico, Peru, and Chili; the emigrants from those mining countries had greater facilities for the time being to reach their destination, and in less than nine months after the discovery there were about 15,000 foreigners on the ground; to these were afterward added adventurers from China and New Holland. At first there were comparatively few Americans, but soon they came, and like a flood. At the end of the year, 1849, as reported by a United States commissioner, there were between 40,000 and 50,000 Americans in the gold region. · The yield of gold during the first two years was $40,-

000,000, and in one year, 1850, the amount reached $50,000,000.

The Location.—The great valley of California consists of the basins of two rivers—one running south, the other north, and uniting in the Bay of San Francisco, through which, by a side cut, they gain admission to the Pacific. These two rivers—the Sacramento and San Joaquin—drain the gold-bearing region of the State. The combined length of their valleys is nearly 500 miles, while their width ranges from fifty to eighty—30,000 square miles. They lie between two mountain-ranges : on the east the Sierra Nevada—the Cascades of Oregon—and on the west, parallel with the ocean, the Coast Range, though of a much less elevation, and not so compact, but broken into a series of high hills. The gold-bearing rocks lie through the length of the combined valleys, and along the western base of the Nevadas in a belt of from forty to fifty miles wide. Nearly the whole width of the Sierra Nevadas from base to base—on an average about seventy miles—is on the western slope, which comes down to 300 feet above the ocean, while on the eastern side the slope is very much steeper, and is only five or six miles wide, but it terminates in the Great Basin, which is itself 4,000 feet above the ocean. At right angles to the Nevadas or nearly so, along their western slope, are interspersed hills with steep sides, and corresponding deep ravines and gorges, through which at certain seasons the waters from the mountains have been, for long series of ages, rushing and tearing away the soil, disintegrating the gold-bearing rocks, and leaving the gold free to sink of its own weight into the alluvium, for the miner to find in the "diggings" known as placers. It is not within the scope of this volume to enter upon a description of these once marvelously rich deposits, but which are now almost exhausted. Geology says: "The auriferous veins of California are parallel to each other and to the Sierra Nevadas, except a few smaller ones. . . . These fissures or veins seem to have been all produced at the same time, when the latter were pushed up." (*Le Conte's Geology, p. 201.*)

Beach-Mining.—As the name indicates, this is done by washing the sands on the ocean-beach. The most profitable portion of the coast for this purpose is that between Cape Mendocino, California, to and beyond Cape Blanco, Oregon. During storms the waves roll high against the shore—the base of the foot-hills of the Coast Range—and wash down the soil which has within it particles of gold. The latter, being thus freed, sinks amid the surging waters, and when the storm subsides, and the tide is out, the sands sparkle with gold. The miners, as time is precious, send early in the morning an experienced and skillful man, on horseback, who, riding as rapidly as possible, ascertains where is the best prospect for gold ; this precaution is always necessary, as that positiou changes in almost every storm. The sand is hastily gathered before the tide comes in, and from it the gold is obtained by successive washings. This species of mining, especially after storms, continues from day to day while the tide is out, until that portion of the beach is explored. The success in this class of mining is quite variable, because on one day the sand will be full of golden specks, but it often happens that by the next it has been washed away, and the bare rocks are alone visible. Companies of miners station themselves in a camp on the bluff, and, after putting the sand in rawhide sacks, it is carried to the place of washing on the backs of mules. (*Hittell's California, p. 317.*)

Placer-Mining.—This was the first method of mining adopted, as being the most simple and inexpensive. Water was used to separate the particles of gold from the alluvial soil in which they were originally washed down from the mountains, and derived from the disintegrated rocks, known as gold-bearing quartz. The abundance of water coming from the Nevadas is of untold advantage to the California placer-miner, compared with the difficulties experienced by the same class in Australia, where the "diggings" are much deeper, and the gold unevenly distributed or "spotted" as the miners say, while the water is quite scanty. The abundance of water led to the adoption of a second method of mining, the hydraulic, in which water is brought in tubes or hose of immense size and strength, and poured with tremendous force against the side-hill or gravel in which grains of gold are hidden. The water, thus violently driven, soon moistens the soil and washes it down, meanwhile liberating the gold, which of its own greater weight sinks to the bottom of the sluices, that are so fitted as to collect it, while carrying away the earthy matter. This method of mining is available where there is a sufficiency of water

Quartz-Mining.—When the placers were about exhausted, and the hydraulic process did not give full satisfaction, capitalists entered upon the business of mining; but they sought the gold-bearing quartz in its home in the mountains, and thus the mining industry gradually became systematized. Now commenced the labor of obtaining the gold-bearing quartz from the vein by drilling and blasting amid the ledges and walls of hard rock. This quartz when conveyed to the mills is subjected to a tremendous pounding by stamps, and it is thus crushed and reduced to powder. The powdered mass is also subjected to washings in sluices, and, while the lighter particles of the quartz are carried away by the current, the gold sinks, and is caught in receptaeles designed for the purpose. Then quicksilver or mercury is mixed with this prepared mass, and, having a strong affinity for gold, speedily attracts it to itself; again water is used, and the remaining earthy matter carried off. In order once for all to free the gold, this cleaner mass is put into a retort, and sufficient heat applied; the quicksilver, rising in fumes, passes through a tube into a chamber where it is condensed and caught in vessels, and is ready to be used again, while the now pure gold is left within the retort.

The Outlook.—The system here described of reducing the golden ore has become an important industry of the Nation. The veins or fissures of gold-bearing quartz pervade the west slope of the Nevadas to such an extent that we should be inclined to limit it only by the size of the mountains themselves. From year to year fresh mines are discovered, and old ones worked deeper and deeper, and the expense is increased in proportion; but that is compensated by the greater scientific skill acquired by those operating the mines and the mills. We can not go further into detail, in respect to the numerous mining districts extending along the entire western slope of the Nevadas, as they are very similar in character. They all appear to be sufficiently rich to remunerate the miners employed and the capitalists therein engaged, and yet it is evident that a dollar earned in the gold regions costs as much exertion as it does in the ordinary forms of industry.

Idaho Gold-Field.—We now come to the west slope of the Rocky Mountains, and commence with Idaho. The Governor of this Territory, in his report (1885) to the Secretary of the Interior, states that there are within it 200 mines paying dividends. The placer-mines on the Shoshonee (vulgarly, Snake) River, are rich and large in extent, but not half their richness has been told or published to the world.

The gold is found in the gold bars or black sands of the river. It is quite difficult to separate the gold from these sands by washing, though many parties make the operations pay largely. These placer-mines only indicate what will no doubt be in time the gold-bearing quartz in the mountain ledges, the original sources of these particles of gold thus freed from the disintegrated rock. The presumption is, that Idaho will yet become a great mining State, when the industry is more fully introduced. The mineral resources of the Territory constitute one of its greatest interests, and, as far as the gold and silver lodes have been worked, the mining industries are in a very prosperous condition.

Rich gold-mines are in the vicinity of Boisé City, while the Cœur d'Alene region has numerous remarkably rich placer-mines. The great Salmon River basin has also a large and productive goldfield, and the Wood River belt of gold and silver districts extend for sixty miles along the streams. There are from time to time new discoveries made and new mines opened, and the product of gold alone was annually about eighty dollars to every inhabitant of the Territory—men, women, and children. Placer-mines are in great numbers all along the valleys of the rivers.

Montana Gold-Field.—Montana is said to be third only to California in the production of gold, while Colorado is the second. The mountains seem to be weather-worn or abraded by storms, and, what is unusual, have beds of gravel within which are placer-mines of gold—that is, on the mountain-tops, instead of being in the valleys, as in California or the foot-hills of the Nevadas. " Veins of gold and silver and copper, and lead, have been found in great numbers in nearly all the explored mountainous portions of the Territory, and placer gold widely distributed." (*Prof G. C. Swallow's Report.*)

Oregon and Washington Gold.—Gold occurs in a number of counties in this State, and has been mined since 1851. In the eastern portion, on the borders of Idaho, a series of mines extends nearly across the State. These ledges of gold and silver appear to belong to the same general field as that of Idaho. Baker City, on a tributary of the Shoshone or Snake River, is virtually the center of the gold-mining industry in that eastern section, comprising the counties of Baker, Grant, and Union, as . the mines are within a radius of *sixty* miles of that city. A gold belt extends also across from Northwestern California into Southwestern Oregon; within this range are a number of mines opened and operated.

According to the Governor's report (1886), valuable mines of gold and silver are known to exist in Washington Territory.

Wyoming Gold-Field.—This Territory has its great mineral wealth in coal rather than in gold and silver (p. 87), yet she has numerous deposits of gold in the usual forms of placer and gulch mines, and no doubt her mountains, when more fully explored, will furnish lodes of gold-bearing quartz. Gold was discovered in 1867 in the southwestern portion of the Territory, and thither flocked great numbers of miners, who established camps in the vicinity of the famous South Pass. A number of these grew into towns, such as Lewiston, Atlantic City, South Pass City, and others; around nearly all are prosperous settlements. Ere long the mining of the lodes or fissure veins amid the granite was introduced and crushing-mills erected. Numerous mines have been opened on Seminoe Mountain belonging to the Sweetwater Range; this is at an elevation of 10,000 feet above the ocean. On the eastern slope of these mountains are a number of gold-mines, which are pronounced by Prof. Aughey, Territorial geologist, to be of a high grade; there are also many indications on the mountain of lodes or veins of gold-bearing quartz. In the product of gold these quartz-rocks vary from ten up to sixty dollars a ton. Wyoming has great facilities in operating stamp-mills, as she has an abundance of coal within reach, and of easy transportation. In the Black Hills, one third of which lie in the northeast of the Territory, is also found gold which has been obtained thus far only by gulch-mining. The gold is more abundant than silver; yet in the Hills it is often found in connection with both silver and copper.

Utah and Colorado Gold.—Utah has a remarkable mining area of different metals—gold, silver, and lead—that is almost coextensive with the great number of the mountains of the Territory. In nearly all the counties, mines of gold have been opened and operated more or less; the placer-diggings have been in the main exhausted, but they were so carelessly worked that, it is said, for the most part, it will pay to work them over again. The lodes of the gold-bearing quartz, however, still remain in the mountains, and the gold waiting to be extracted by means of more scientific methods. The gold production of the Territory is worth only about one fourth part that of the silver. The latter metal is found in combination with lead in a manner similar to that in which it occurs in the famous Leadville mines in Colorado; hence the output of lead in Utah is quite large, sometimes amounting annually to more than twenty tons. Colorado and Utah being adjacent, partake of similar characteristics in their mineral wealth, and, though Colorado is second to California in the production of gold, yet her annual output of silver is about five times the value of her gold.

Arizona and New Mexico Gold.—In these Territories gold is distributed in very numerous places, and there is evidence that many of the placer-diggings were worked, as well as some deep mining, ages ago by the Spaniards, who in some instances appear to have held the Indians as slaves, and compelled them to labor in the mines. The placers in some places were no doubt rich; but the skill of the Spaniards of that age did not rise above the crudest form of mining. As yet there has not been very extensively introduced mining by scientific methods; these two Territories are richly endowed with the precious metals—more of silver than of gold—as well as with the inferior and more useful.

Alaska Gold.—This Territory was purchased in 1867 from Russia, by the United States Government, for the sum of $7,200,000. Its gold-mines have attracted attention, and for a number of years placer-mining has been carried on along the valley of the Stikine River, and with profit. A number of gold-bearing quartz-lodes have been found on Baranoff Island, in the vicinity of Sitka, the capital of the Territory. An extensive mill and works have been established on the island to crush the quartz from these lodes; the mill when in operation runs one hundred and twenty stamps and has forty-eight concentrators. The amount of quartz crushed daily is about three hundred tons, and it assays from eight to twenty dollars a ton. Gold is known to exist in the Coast Mountains and on some of the islands —as on Douglas Island—having the same geological formation as that of the mainland; and also, in a number of places in the interior, as reported by Lieutenant Stoney, of the United States Army, who has been conducting explorations within the Territory. (*Governor's Report, 1885.*)

XVIII.

SILVER is found more than any other metal in combinations with different substances, as sulphurets or oxides, etc., and with other minerals; it also occurs in several geological formations, that are far separated in age and in location: hence it appears to be universally diffused. Metallurgists have proved by experiment that, of all the numerous ores found in Nature, only one in *seventeen* is free from the presence of silver. Minute traces of it even exist in the waters of the ocean and in organic substances. It is often found more in connection with gold and lead, especially with the latter, than with other metals, and, from its combination with these two, we have generally derived our silver. The affinity of silver is so strong for other substances that it becomes difficult to treat of it separately from these alliances. Its ores likewise are frequently imbedded in fissures or cracks of the oldest rocks; and the veins within these fissures have often been worked to great depths, but without the richness of the ore diminishing. The main source whence silver has been derived, for two hundred years after its discovery, was Mexico and South America, and its passing into trade as a medium of exchange influenced the industries and commerce of the civilized world, as in more recent times the influx of gold from California produced a similar effect.

Nevada Silver.—On the western slope of the Sierra Nevada we have seen immense gold deposits, either as placer or as gold-bearing quartz, and now nearly opposite the center of this goldfield on the eastern slope of the same mountains, and on the west side of the Great Basin, we have correspondingly large and numerous deposits of silver-ore. Until these were discovered, it was thought that silver, except in small quantities in combination with the ores of other metals, as lead or copper, or as we have seen in connection with gold, was nowhere in existence within the United States. But here, contrary to precedent and analogy, were immense ore deposits in which silver itself predominated.

The Comstock Lode.—The center of this mining district—the Washoe—is Virginia City, which is perched on a rocky shelf on the eastern face of Mount Davidson, more than 6,000 feet above the ocean, the mountain itself rising 2,000 feet higher. On this spot, in 1860, stood a single log-hut. The year previous two prospectors for gold—named Comstock and Jenrod—near the shelf discovered ac-

13

cidentally what to them seemed a " vein of dark ore "
of some kind. Not knowing its character, they sent
a specimen to San Francisco for assay. It was ascertained to be very rich silver-bearing quartz. This
deposit of silver-ore proved the richest in that
region. The fact becoming known, miners flocked
thither in multitudes, and soon the whole mountain
was marked out in " claims "; and the largest vein
was called the Comstock, in honor of the discoverer.
Ere long, nearly one hundred companies were operating these claims. The latter varied from 25 feet
to 200—each company being entitled to the whole
depth and width of the vein or lode, whatever that
might be.

Owing to the very great inclination, almost perpendicular, of the dip of the ore amid the hard rocks,
the labor of working the lodes is much increased,
and the expense correspondingly great. These
mines produce gold as well as silver, but a greater
proportion of the latter. The Comstock lode has
been the most productive, not only because of its
richness, but also because of the numerous companies engaged in working it. The veins of silver-
bearing quartz run along the eastern slope of Mount
Davidson in parallel belts, nearly 1,000 feet long, and
they have been explored nearly four miles in different directions, by the underground workings, and,
as the veins run down, they can in some cases be
reached by horizontal openings in the side of the
mountain. Within the subterranean region at one

time there was no intermission of labor, that being kept continuous by relays of miners : here was no Sabbath, no winter, no summer; thousands of burners made it as light as day, while there was scarcely any variation in the temperature. As proof of the energy displayed, these mines have produced in twenty-five years about $300,000,000 in gold and silver bullion.

Sutro Tunnel.—Owing to the downward direction of the veins of ore, the workings of the Comstock mines soon became very deep, and only at great expense could they be ventilated or freed from water, and the ore raised to the surface. To obviate these rapidly increasing difficulties, the " Sutro Tunnel"—named from the projector—was designed to drain and ventilate the several mines and take the ore out on a level. This tunnel is fourteen feet wide and twelve feet high, with a double-track tramway by which the ore is taken out, and drains on either side draw off the water, while the mines are also kept cool and ventilated. It enters the mountain about 2,000 feet below Virginia City, and is nearly five miles long, and from it extend galleries in every direction in which the lodes or veins lead. There are many other localities, such as Eureka, where lodes are worked. Nevada, on an average, produces three times as much silver as gold.

Liberal Mining Laws.—The mining laws of the National Government are very liberal to the miner, as it looks upon him as a *poor man ;* these laws are not strictly just to the Nation itself, the latter being the real owner of the public lands. These laws permit the miners to go upon the public domain and without restriction to work the mines, the Government not demanding a royalty or percentage on the minerals thus obtained, as *is customary with other governments.* The miner marks out his claim, and it is protected for him by the law. This is a counterpart of the still more benevolent and humane *homestead law,* so comprehensive in its beneficial effects upon the people at large, as it applies especially to the family and all its members.

Leadville Silver-Mines.—In the mining districts of the State of Nevada we have silver in abundance, but in combination with gold. We now pass directly east for 753 miles to Leadville, in the State of Colorado, in the vicinity of which place we find silver, but in combination with lead. The former mines are on the east slope of the Nevadas, and the latter amid the Rocky Mountains. The characteristics of the lodes or veins are distinctive: in the former they are almost perpendicular, running down toward the base of the mountain; in the latter the mineral deposits of lead and silver lie comparatively horizontal. In consequence of this difference, the mining operations in the former, as we have seen, are very expensive, while in the latter the cost is much less. In Colorado more mines, in proportion, have been opened by individuals, and by companies having comparatively small capital, than in the mines in Nevada. The great production of silver in the United

States, within recent years, has been principally derived from these two mining regions, nearly 800
miles part—the one on the east slopes of the Nevadas,
the other amid the Rockies. The Director of the
Mint estimated the production in the United States
of the precious metals, in 1885, to be of gold, $31,-
801,000, and of silver $51,600,000. It may be noted
that a large amount of gold and silver bullion, mined
by individuals, does not reach the Mint, but is distributed in various ways for ornamentation or in the
arts.

Arizona and New Mexico Silver.—These two
Territories have the reputation of possessing an unusual share of barren lands, and of regions destitute
of useful vegetation; but to-day the indications are
that these disadvantages are compensated to the people by means of the vast amount of mineral wealth
existing within their rough mountains and in the soil
of their sterile plains. In these two Territories silver
appears to be found in combination with gold, copper, and lead. In other places, the process of obtaining silver is incidental, rather than otherwise, to
the mining of the metal, with which it happens to be
combined; but in two instances, at least, in Arizona
and New Mexico, silver predominates to such a degree that it is mined directly for itself.

About a hundred and fifty years ago, advent-
urous priests brought to Mexico rumors of the
marvelous wealth of native silver that existed in
the now Arizona Mountains. They said it was found

in great masses, weighing hundreds of pounds. These reports induced multitudes of Spanish miners to flock to these regions; but, after more than three fourths of a century, the Indians, whom they had forced to work in the mines, rose in rebellion and drove off the intruders, priests and all. The remnants of tunnels and shafts in many places show where these pioneer miners of the white race once worked; these remains are in the vicinity of the "Old Missions," now in ruins, and here also are seen the slag and the *débris* of their crude furnaces. Notwithstanding these drawbacks, the richness of the ore must have made them quite successful. When Arizona passed into the possession of the United States (1848), not a single mine was in operation within that region; they had all been abandoned for nearly a third of a century, and the irrepressible Apache roamed over its plains and mountains unmolested.

Old Mines reopened.—Rumors concerning the "Mission" mines induced American miners to prospect the southern portion of Arizona, and they revealed to the world the immense measures, especially of silver, that were stored in the "land of sunshine." In the region south of Tucson, amid the Santa Rita Mountains, were found the relics of the old mines; these were soon reopened and worked with success. Meanwhile, discoveries were made of numerous rich deposits of silver combined with other metals, but frequently in an almost pure state.

The headquarters of these active mining operations were at Tubac ; but reduction-buildings were put up in many other places, the latter becoming centers of their respective districts. " The ores were exceedingly rich and easily reduced," and the work of development went steadily on, though the miners were constantly harassed by the hostile Apaches. In the northern portion of the Territory, in Yavapai and Mohave Counties, have been discovered ledges in which are gold, silver, and copper, while rich placer deposits prevail extensively over that region, making it a very valuable mining section. These deposits are south and east of the Colorado, and on tributaries of that river.

Tombstone.—A brief notice of Arizona's richest mining town may not be uninteresting to the reader. It was inferred from vague rumors that rich deposits of silver existed amid the western foot-hills of the Dragoon Mountains, which stand in a southeast direction from Tucson, in what is now Cochise County ; but that entire region was a favorite resort of a portion of the Apache Indians, under the leadership of a noted chief (Cochise), who killed every prospector he discovered within his domain. Cochise had his main fortress in these mountains, and his faithful lookouts on their crags.

Mr. A. E. Shiefflin was a persevering prospector, and withal very bold, prudent, and determined. He resolved to penetrate the mystery, and "prospect" that region ; his friends endeavored to dissuade him, prophesying that he would find nothing more than his *tombstone*. But undaunted, he set out, and, avoiding the hostile chief, cautiously explored the region, and some six weeks later he returned (February, 1878), and announced that he had discovered the silver deposits, which have since become so famous under the name of TOMBSTONE, which name Shiefflin gave to the district or camp that has since grown into a mining town. The news of this grand, discovery spread far and wide, and the miners, ever restless, regardless of the hostility of Cochise, flocked to Tombstone, coming in numbers from the Pacific coast, from the East, and from other mining camps. The high grade of the ores, the immense deposits, the easy mining and comparatively small expense of reducing the ores, invited capital, and the great yield of bullion made Tombstone famous as the leading mining town of Arizona. Reduction-works and stamp-mills soon sprang into existence, with every appliance for obtaining the silver. The ore was first taken out almost as from a simple quarry ; but the lodes or veins extend downward, and to follow them shafts have been introduced. The mineral belt of Tombstone extends about eight miles east and west, and toward the south about twenty-five, having an area of nearly 200 square miles. Upon the whole, Arizona appears to be pervaded to a very great extent by rich deposits of minerals of almost every kind.

Lake Valley.—In an easterly direction, and distant about 200 miles from Tombstone, is another remarkable deposit of silver-ore. It is located in Sierra County, Southwestern New Mexico, in the Lake Valley mineral district. The amount of mineral wealth in this region would appear incredible if the statements made in respect to it were not authenticated by gentlemen whose judgment and experience command respect. The ores found are of different degrees of richness, so that here " ores that run from $200 to $300 to the ton are classed as of low grade." In a paper prepared on this subject by S. H. Newman, Esq., this assertion is made: " We believe we saw in the mines not less than $15,000,000 worth of ore." This statement Mr. William G. Ritch, author of " New Mexico Illustrated " (pp. 42, 211), verifies. The ore crops above-ground and continues for a depth of fifty feet, and along the hill-side for a distance of about two fifths of a mile. The veins run downward, and shafts have been introduced, the excavations extending under a ledge of limestone some six feet in thickness; the latter, supported by timber, serves as a roof. Here are masses of silver-ore piled one upon another, as if forced into their position by some internal convulsion. The ore is of different shades of color, as shown by torchlight: sometimes the rock is dark, like the slag from an iron-furnace; then often of a reddish cast, and again similar in appearance to an amalgam of quicksilver; in another place " the ore hangs in beautiful, glistening, soft chloride crystals which feel damp in the hand, and when compressed yield to the pressure and assume the shape of the closed palm, like dough." The latter is easily smelted, " the flame of the candle sending the virgin silver dripping down the wall like shot. . . . These chlorides run about $27,000 to the ton." One lump of ore, a cube about three feet square, and valued by experts at $7,000, was taken from this mine and exhibited at the general exposition of the mineral products of the Rocky Mountain States and Territories, held at Denver in 1883.

Other **Mines of Silver-Ore.**—The deposits of silver in New Mexico are by no means limited to the Lake Valley district, as there are numbers of such in the region round about. They are all in the valley of the Rio Grande. This mineral belt extends around for a number of miles, and the presumption is that, especially toward the north portion of the Territory, there are other deposits of silver-ore equally rich, that will in turn be discovered. The indications are, also, that all the valleys between, and the foot-hills of the subordinate ranges of mountains in the western part of New Mexico, and the bordering eastern portion of Arizona, are permeated more or less with lodes or veins of minerals, forming a sort of network in both Territories, and that some districts hitherto deemed barren wastes are found to be rich in minerals; while, as we shall see, there is much more land in these two Territories capable of being utilized for crops of various kinds, and for stock-raising, than has been generally supposed.

Utah Silver.—Utah ranked *third* in the production of silver in 1885, Colorado and Montana being the *first* and *second.* (*Mineral Resources of U. S., 1885, p. 201.*) The mines of this Territory belong to the great silver field that occupies so large a portion of New Mexico, Arizona, and Colorado, but its ore partakes more of the characteristics of that of the last, inasmuch as the silver is found in combination with lead, while the great deposit in Nevada is in connection with gold. The ores of silver are scattered over the Territory, and in consequence there is an equal diffusion of mining establishments, some of which are very extensive and very perfect in their appliances for obtaining the ore from the mines by means of shafts, etc., as well as of smelting the ores, and separating the lead from the silver. Mining these metals has become a prominent industry of the Territory.

Idaho, Montana, and Wyoming Silver.—We pass now to the north. In the mines of Idaho, silver is found in connection with gold, as in the remarkable mineral belt extending along Wood River for sixty miles—the silver running from 100 to 350 ounces to the ton of quartz-rock ore. The Sawtooth and Lava districts have extensive and rich silver deposits, but are not yet fully developed. (*The Governor's Report, 1885.*) Montana is, in the production of silver, according to the " United States Mineral Resources " (1885), second to Colorado.

In Wyoming Territory, gold is more prominent

than silver. In one instance, the silver is in connection with copper, while the strata immediately below show veins of native silver, and still lower veins of silver and gold combined. In the subordinate Laramie range of mountains is a distinctively copper-bearing region, its ores having more or less of silver.

Pacific Coast Silver.—The " Mineral Resources of the United States " does not mention silver as one of the metals belonging to California, though there may be hidden deposits. In the eastern counties of Oregon, silver is sometimes found in connection with gold, and in mining the latter it is often obtained, but in limited quantities. The Governor of Washington Territory, in his report (1886), enumerates silver among the valuable metals of the Territory ; but we would infer that it is found incidentally and only in small quantities.

" One third of all the gold and one half of all the silver annually produced in the world are supplied by the mines of the United States." (*Mineral Resources of U. S., 1883, p. 185.*)

XIX.

WE have seen, in the case of iron, how often, near by the deposits of its ore, were found the coal and limestone so essential in smelting it. We now notice an instance in the economy of Nature similar and equally striking, that of the proximity to mines of gold and silver, of cinnabar, the ore from which quicksilver is derived. The latter is as effective in separating the precious metal from the gold-bearing quartz as are the coal and limestone in obtaining iron from the ore. Thus, within available distance of the gold and silver mines of California and Nevada, is found cinnabar, or the red sulphuret of mercury, stored away in the strata of the earth, to be utilized when needed.

New Almaden.—Inside the Coast Range, twelve or more miles southeast of San José, are said to be the richest mines of cinnabar in the world. These are the New Almaden, named from similar mines in Spain. In the vicinity and State are other deposits, as Fresno, the New Idria—named from an old mine in Austria, that has been worked for three hundred years—

Napa, and others, but they are all inferior in richness when compared with the first mentioned. The cinnabar at New Almaden is imbedded within walls of flint and slaty rocks; it is found in a series of beds and layers, so that the workings are very irregular. The masses of ore are separated by strata of rock, which are variable in thickness, while the crevices between them are often filled with seams and bunches of ore. The latter, thus disconnected, is liable, as the miners say, " to nip out," and suddenly the mine is found to be exhausted. In addition, when the cinnabar is worked out in one place, no clew is given by which to find another. Nature seems in this case to have put all rules and analogies at defiance, and the mine found has been stumbled upon by accident.

The Operation—how conducted.—The veins of ore run down into the earth, and are mined by means of a perpendicular shaft, several hundred feet deep. The miners have followed the veins thousands of feet by means of galleries opened right and left from the shaft and from one another. At first the miners were a motley company : here were English from the copper-mines in Cornwall, Welsh, Scotch, Irish, and Mexicans, the latter being more skillful in this kind of mining. The labor is performed in quite a primitive manner : the ore is brought up the shaft in large buckets by aid of the windlass ; in the same way the men are passed down and up, to and from their work.

The ore or cinnabar is of a reddish color, similar

to that of ordinary bricks. It is first prepared by being crushed and fashioned into blocks, which are arranged with spaces between them in ovens or kilns, where they are subjected to heat so intense as to become white, the flame passing between the blocks. The heat first causes the quicksilver to exude from the earthy matter, and then by its intensity changes it into the form of vapor, and when in that state it is conducted by tubes into a chamber designed for its reception. In the latter, when becoming cool, it condenses on the walls and trickles down into channels arranged to convey it to a reservoir, from which the iron flasks—glass could not bear the weight—are filled. The quicksilver, thus pure, is ready for use. As water, when passing off in vapor, leaves all the ingredients which it held in solution, and rises perfectly pure, so quicksilver, under the influence of intense heat, rises in vapor and frees itself from all extraneous matter.

The Effects produced on **Gold-Mining.**—The discovery and successful working of the New Almaden had a great influence on the mining industries of the adjoining States, since quicksilver is universally used in the extraction of both gold and silver from the ore, and also being furnished in abundance and at a cheap rate. The peculiarity that quicksilver has of being used over and over again, in the extraction of gold and silver, and the comparatively small amount required in the arts or ordinary industries, do not demand that it should be supplied in a superabundant degree; its production in Nature is quite limited. In addition to California, it is found only in Spain and Austria in the Old World, and in Peru and Chili in the New.

The trade of the world in this metal is virtually controlled by Old Almaden in Spain, in connection with New Almaden in California. Spain supplies Europe and partially Mexico, and California the United States and China, and also makes large exports to Mexico. At one time China was supplied entirely by European dealers; but New Almaden, having on hand a large surplus, suddenly, in 1879, with nearly 37,000 flasks, appeared in China and overstocked the market, and at a much lower rate. The dealers, finding they could not compete with this unlooked-for rival, packed up their entire stock and carried it back to Europe, and, as far as known, have never yet returned. In 1885 New Almaden produced 21,400 flasks, while all the other quicksilvermines of California furnished 10,673.

XX.

THIS metal has been known to man from remote ages, and was perhaps the first with which he became acquainted. Copper, in some form, has in its use always preceded iron—the latter being so much harder to refine, and the only metal whose pieces can be welded together; but copper can be so heated as to melt and become even a fluid. Tin and zinc also can be reduced to a fluid state by heat, and the workman is able to combine either with copper, when both are in that condition; thus, in combining in certain proportions with tin, he produces bronze, and with zinc under similar conditions he produces brass. The ancients had means —to us unknown—of hardening bronze so as from it to make instruments with which they could cut granite. Copper in its different combinations is coming more and more into use in diversified forms, as science discovers the numberless applications that can be made of it in the various industries of the day. Copper when pure is nearly red in color; it is exceedingly tough, and can be drawn out in wire or made into sheets.

The Universality of Copper.—Copper, as found in Nature, is diffused as much as silver, if not more, and its natural combinations are nearly as numerous. It exists in many soils, in sea-water, and is traced in plants and animals. With the exception of gold, copper is found in a pure state more than any other metal; and nowhere do such immense masses of pure copper exist as in the United States, it being found in twenty-one States and Territories. When pure, it is also found in small lumps—but in a crystallized state, the usual form being octahedral, or eight-sided. The ores of copper are usually very beautiful, having different shades of color, and each one brilliant, as red, blue, green, etc. Its natural combinations with other substances in the form of ore are so numerous and so diversified, that to go into detail would be out of place in this sketch; of these we therefore notice only one, that with sulphur. The common yellow copper-ore is defined by Whitney as "copper pyrites [a sulphuret] mixed with the sulphuret of iron." This is the class of ore which furnishes nearly all the copper that is derived from the mines of Cornwall, England; and indeed, by good authority, this class is also estimated to furnish two thirds of the copper mined in the world.

Copper on the Atlantic Slope.—Copper-ores are found, but in small deposits, along the Atlantic slope from Maine clear round to the southwest side of the Alleghanies in East Tennessee. These deposits are very similar in character; they never occur in what geologists call transverse or fissure veins, and though these limited masses of ore are parallel with the formation, geologically speaking, the veins lie separate from one another. The ores thus situated are nearly all the usual "copper pyrites." Though the localities where copper is found are numerous, the limited amount of ore scarcely remunerated the miner, and then only when the demand for domestic use was great, and in consequence the American people were forced to depend nearly altogether upon the foreign supply. It may be proper to remark that copper, as a natural resource, is to-day about exhausted in the Eastern portion of the United States.

In the State of Maine are a few "quartz-veins" of copper-ore. Isolated mines having similar characteristics exist also in Franconia, New Hampshire, but none have been worked to any extent. Several places in Vermont are named as having deposits of copper pyrites; and others where is found a "green carbonate with particles of vitreous ore"; while, in Massachusetts, copper-ore occurs in the vicinity of Northampton, but in small quantities and in connection with lead. Connecticut has, near Bristol, the most extensive deposit in that section of the country. The ore at the surface is mostly variegated copper, but there is no evidence of a fissure-vein. This mine has paid the operators better than any other one in New England, the ore yielding about 32 per cent of copper. The mine has been opened by shafts; but neither the amount of ore nor its richness has warranted extensive outlay for mining purposes.

Copper has also been discovered in a number of places in the States of New York, New Jersey, and Pennsylvania, but in such small quantities that the mines have long since been virtually abandoned; likewise in the States of Maryland and Virginia, but they were never of much value. These mines for a time furnished a limited amount of copper for domestic use.

North Carolina, Tennessee, and Georgia Copper. —North Carolina has, apparently, more copper-ore within her boundaries than any State east of the Alleghanies. Dr. Genth, mineralist, says in the "Handbook" of the State, 1886: "Copper-ores have been found in many localities throughout the State, . . . the principal ore being copper pyrites [sulphuret]; and there is every reason to believe that many of the mines require only a fuller development to enable them to furnish large quantities of valuable ore. . . . Almost all the copper-mines in the central counties of the State have been first worked for gold. . . . The general character of these mines is that the so-called brown gold-ores are replaced by quartz richly charged with iron pyrites [sulphurets], more or less mixed with copper pyrites, the latter increasing as the mine deepens, and in many places becoming the only or the predominating ore, and forming a regular copper-vein." These mines seem to be held in reserve for the future, as they are worked to little extent at present.

In Polk County, in the extreme southeastern portion of the State of Tennessee, is a district in which occur deposits of copper-ore, which, the "Handbook" of the State says, "is capable of being a great source of wealth." Some of the veins have been explored to a depth of 200 feet, and give evidence that they extend still farther. The ores near the surface are red and black oxides, but at a greater depth these give place to a sulphuret. The greater facilities in obtaining copper in the Lake Superior region had an influence in diminishing the mining operations that once prevailed in this district. Copper-ores are also found in other places in the vicinity, all of which in due time will be utilized, when they become accessible to more efficient fuel for smelting, and also to the markets of the Union.

Copper-ores occur, but in limited quantities, in several counties in the State of Georgia. These veins in numerous places can be traced for many miles; in one instance a deposit exists on the top of the Blue Ridge in Lumpkin County. The outcrop of the copper can be traced for several miles, it following the northeast-southwest trend of the mountain.

Lake Superior Copper.—We come now to the great copper region of the United States, or of the world, occupying a portion of the south shore of Lake Superior, and islands within the lake itself. The mines are located principally within the three counties of Keweenaw, Houghton, Ontonagon, and the island and county of Isle Royale. These deposits of native copper became known to the civilized world through French Jesuit fathers, in the latter portion of the seventeenth century. They explored the entire region and made known its mineral wealth, but in somewhat exaggerated terms, as future investigation proved. They proclaimed that there was copper, and also gold and precious stones; the latter were simply agates.

A little more than one hundred years ago coppermining began on the Ontonagon River, about forty miles from its mouth, in a district full of mines today; here the fathers saw the famous mass—estimated at one hundred tons—of native copper lying , on the west branch of the Ontonagon. About half a century later the National Government sent army officers to explore the country round the Great Lakes; though their orders were only partially carried out, their report revealed the fact that here were immense deposits of copper and of iron ores.

The Copper-Fever.—When the region around Lake Superior was opened for settlement (1844) by act of Congress, intimations had already spread far and wide of the untold wealth of minerals—of iron and copper, especially the latter—that lay around the shores of the great lake. Specimens of native or pure copper were exhibited in the Eastern cities, and the story was told of the great masses of the same that were to be seen on the surface, while the veins of ore were numerous and enormous. Then began what is known as the "copper-fever of 1845," when thousands flocked to the land of prospective wealth, going first to that long and narrow peninsula, Keweenaw Point, the shores of which were whitened with the tents of miners and speculators.

An official in a department at Washington had issued "permits" for persons who wished "to select and locate on tracts of land for mining purposes." There were 375 such leases granted on that slender point alone. The selections were made for the most part at random, the selector having no knowledge whatever of mineral lands—numbers being made on rocks that were barren of copper or of any metal. Speculation raged, and for two years stocks in these enterprises ruled high. The bubble burst, and thousands were financially ruined, the climax coming with a crash when the issuance of "permits" was suspended by Congress because it was illegal. The latter body ordered a thorough survey of the mineral region, and also published accurate and authorized maps of the entire districts. The basis being changed, in a few years mining of copper was placed on a safe foundation, and as such has continued a prominent industry.

The Mound-Builders as Miners.—Long ages ago people of a different race left traces of their rude mining operations, both on Keweenaw Point and along the south shore, and on Isle Royale, which lies in the lake some sixty or seventy miles directly north. These ancient excavations were of various depths, in one instance reaching fifty feet in the solid rock. In one place on the island a series of these mining-pits extends for nearly two miles, having an average width of 400 feet and a depth of twenty, the miners taking out of the solid rock the belt of copper. The tools of these ancient workmen, thousands of which they left behind, consisted of stone hammers, shaped somewhat like a modern axe, but having round the head quite a deep groove within which, to serve as a handle, the green withe was skillfully wound. Here are also found pieces of charred wood or fire-brands, that had resisted the aggressions of time, giving evidence that these miners also used heat in detaching the copper from the rock—the latter probably becoming cracked by being first subjected to heat, and then suddenly cooled by the application of water.

The present race of Indians, at the advent of Europeans, had no knowledge of mining, nor had they traditions of the people who made these excavations. The presumption is that the latter were the mound-builders, whose monuments we find on the fertile plains in the Great Valley, and who only visited these regions in order to obtain copper. This we infer from the fact that they left not a vestige of a dwelling, a burial-place, or a skeleton, and that pure copper, either in the form of ornaments or arrow-points, is often found in their mounds. The mound-builders were much more civilized than the ancestors of our Indians, but they were, no doubt, driven by the latter savages from their peaceful homes and forced to seek shelter in the far South, where they evidently became the ancestors of the peoples of Mexico, Central and South America.

The Process of Mining.—The mineral region around Lake Superior is peculiar, inasmuch as the copper is found native or almost absolutely pure, instead of being in the form of ore, and thus requiring to be smelted, but, on the contrary, the process of refining consists more in removing its extraneous surroundings. In addition, the veins within the rocks, that are the most productive, carry exclusively native copper with a small amount of native silver ; and this characteristic continues at every depth yet reached. " The width of the productive veins is usually from one to three feet, they sometimes widen out to ten feet or even more, but rarely continue to hold these dimensions for any considerable distance. The wider the vein, as a general rule, the richer is its metallic contents. . . . The copper is found imbedded in the vein-stone in pieces of every size, from almost microscopic particles up to masses of one to two hundred tons weight." (*Whitney's Metallic Wealth, etc., pp. 259, 260.*)

When a mass of copper, perhaps of several feet in length and breadth, is met with in the vein, the adjacent rock on one side is removed ; this leaves room for the mass to be shoved, and after other preparations heavy charges of gunpowder are placed behind the mass, which when it exploded moves it out of place. Copper can not be broken into fragments by any force of explosion; it only *tears.* The block of copper thus loosened must be divided into pieces of a size convenient to be brought to the bottom of the shaft and thence hoisted to the surface. To make these divisions, hard steel chisels of different lengths are used, according to the size of the block of copper to be cut; to manage this cutting requires at least two workmen, one to hold and direct the chisel, the other to drive it home by the blows of a sledge-hammer. This process is both laborious and tedious; and in addition the fragments of rock or *débris* must be removed out of the way, either by being carried to the surface or transferred to a vacant place, whence the copper has been already taken out. Copper is also often found imbedded in the vein-stone in smaller pieces weighing usually a pound or more. These pieces, when cleaned of the adhering rock, are put in strong barrels and sent to the surface. This class of ore the miners designate "barrel-copper."

The Two Mines.—The veins of copper in this region are imbedded in the rock, and always have a dip or inclination usually about 39°, and this direction being known, enables the miner by means of a perpendicular shaft to strike the vein at any point he may choose. Frequently a number of shafts are sunk, in order to reach the same vein at different points. From the bottom of the shaft he runs parallels in different directions, and by removing, as already described, a sufficient quantity of the adjacent rock, he is able to take out the copper. The mines, in the main, are operated alike, and the process in one illustrates the whole. There are great differences in the richness of the deposits or veins, and in consequence there are often failures as well as successes. Of the latter class are two remarkable instances in Houghton County—the Tamarack and the Calumet—both operating in different portions of the same vein. They are both well equipped, the former having a shaft of the largest size; " its inside dimensions being seventeen feet eight inches by seven feet. It is divided into three compartments, two of which are used for hoisting, and the third for pump and ladder way. . . . This shaft struck the lode at a perpendicular depth of 2,260 feet." These fully equipped companies can bring, each, to the surface daily between 2,000 and 3,000 tons of rock from a depth of 4,500 feet on the vein. The Commissioner on Mineral Statistics says of these mines, in his report for 1884 (page 67): " A rough estimate, based on the most conservative data, makes the total reserves equal to twenty years' work, at the present rate of production of 20,000 tons of ingot copper per annum. This mine on the lower levels is so dry that the water used in the drilling operations is supplied from the outside."

Isle Royale.—Among the numerous islands in the lake, one deserves a passing notice—the Isle

Royale. It is forty-five miles in length, and on an average twelve in breadth. It rises in bluffs sometimes 300 feet high above the water, with here and there a harbor, a low place or cut into the cliff. One of these inlets, Rock Harbor, is more than ten miles long, and only from one quarter to one half a mile wide. At the entrance are a number of beautiful small islands; the former is bordered on each shore by terraces of magnificent evergreens, so regular and so arranged that they seem to have been planted in parks by the hand of man. This island is exceedingly rich in copper, found here almost pure, and sometimes in masses containing hundreds of pounds up to hundreds of tons. A great number of companies are engaged in mining copper on the island: some of these are very extensive. In the State of Michigan there were recently sixty-six corporations engaged in mining copper.

Wisconsin, Missouri, and Texas Copper.—Copper occurs in the State of Wisconsin, but not of sufficient richness and quantity to warrant mining it to much extent. These ores are found for the most part near Mineral Point. The same may be said of the limited deposits of copper within the State of Missouri, as the ore, though it occurs in various places, and in connection with the ores of other metals, is not sufficient in quantity to induce mining operations to a very great extent.

The copper-ores of Texas thus far discovered are located in three fields: The *first*, in which the ores are abundant, is in the valley of the Colorado, on two of its tributaries, Llano and San Saba, and in counties of the same name. These mines give evidence of having been operated by the early Spanish colonists. The copper veins are in connection with granite and quartz, and are generally carbonates, but as they extend downward the copper becomes allied with sulphur and with a minute quantity of gold and silver. These outcrops are quite extensive, the belt extending into the six neighboring counties. The *second* field lies almost north of the first, on the borders of the Indian Territory, in the valley of the Wichita, a tributary of Red River, and in the vicinity of the Wichita Mountains, from which ores are supposed to have been washed down. " Large masses, weighing as much as *two* tons, have been found, but no vein has yet been exposed." (*Mineral Resources of U. S., p. 343.*) The *third* field is in the southwest portion of the State, on the eastern slope of the Chinate Mountains and in the ranges between the latter and Mexico, and also in the Guadalupe Mountains, between the Pecos River and the Rio Grande in the vicinity of Fort Davis. The evidence is presumptive that there are vast deposits of copper-ores, lying in a line north and south, all through the western portion of Texas.

Copper in the Territories.—The copper-ores of Wyoming are carbonates and oxides; they are rich, producing a good grade of metal, and are found in great abundance. (*Governor s Report, 1885.*) But as yet comparatively few mines have been opened; several lodes of copper occur in the region north of Fort Laramie; these contain an ore rich in quality. Southwest of the latter a number of rich deposits have been discovered in the Platte Cañon district, to smelt the ores of which works have been established at the mouth of the cañon on the bank of the North Platte River. The ore is obtained by means of shafts. Throughout much of the southern and middle portion of the Territory, copper-ore is found in many localities, and in addition there is a large area of the Territory that has not been examined for the purpose of finding copper. In a locality northwest of Cheyenne, some twenty-five miles, and on the eastern slope of the Laramie range, is a mining district known as the Silver Crown; here is a deposit of copper-ore, that is mingled with gold and silver. A test mine showed veins of ore from two to three feet thick; the dip of these veins is almost perpendicular. "One of the veins showed 30 per cent of copper, $10.36 per ton of silver, and $10.33 per ton of gold." In another instance the assay gave 46 per cent of copper, and $20.25 worth of silver per ton; and still another assay, from ore selected, gave 57.4 per cent of copper.

Upon the whole, the copper-ores of Wyoming are uniformly rich, and the quantity within the Territory thus far discovered indicates that there are other deposits equally rich and extensive.

Dakota Copper-Ore.—This Territory has deposits of copper-ore, but they have not as yet been developed. (*Mineral Resources of U. S., 1883–1884.*)

Montana Copper-Ores.—Montana is noted for her valuable coal-helds, and she is equally fortunate in the possession of copper-ores, as the latter are rich in character and great in extent. The mining district of the Territory that holds the first place in the production of copper is located about thirty miles south of Helena, in the vicinity of Butte City. In the output of copper the mines here are second only to those of Lake Superior ; in addition, this district produces not only copper, but in connection with it a fair amount of silver. These mines are confined to an area of about two and a half miles in length by about one in width; but within this comparatively small space is concentrated an immense amount of copper-ore mingled with silver. The great size of the Butte City copper veins is remarkable. " Often a vein is thirty feet in width for several hundred feet of its course, and filled with ore the entire distance, and very often these veins show no diminution in the richness of the ore."

The copper deposits of the Territory arc by no means limited to such small quarters, as they are found extensively in many other localities. " These all, without exception, contain copper and silver, though in widely varying proportions, and most of them contain traces of gold." So much is this the case that often mines, which to-day are worked conjointly to obtain copper and silver, were worked originally for silver alone. Then it is proper to take into consideration that these veins have sometimes also "a silver-bearing lead of great value, carrying on an average two thirds of an ounce of silver for each per cent of copper." The indications are that copper-ores abound in the greater portion of the Territory.

Idaho Copper.—In Idaho, according to the Governor's report (1885), are rich and quite extensive copper-mines. They are in different localities, among others a group of mines known as the Peacock; one of these, an immense deposit of ore, lies a few miles from Snake River, in the singular position of being on a mountain 4,000 feet above the river. "The ore runs high in copper, and carries also a high per cent of silver." These mines, from their nearness to the railway, and their facilities for working, will ere long be noted for their output of copper. There are also copper-mines on the Middle Weiser River, which, since the construction of the Oregon Short Line Railway, have become accessible. In addition, large lodes of copper-ores have been discovered on Lost River, east of Idaho City, the ores of which are represented as "marvelously rich." In this vicinity several mines are in operation.

New Mexico.—Near the middle of this Territory, in the valley of the Rio Grande, are large deposits of copper-ore, which consist of the usual kinds or grades that require smelting. As quoted in *New Mexico Illustrated, p. 103*, from a writer in the *Albuquerque*

Journal, the copper-fields in Bernalillo County are described as very rich in ore and of immense quantity, but as yet in trust for the future, and waiting to be developed. In Lincoln County, in the valley of the Pecos, "an immense copper belt has recently been discovered, bearing both native copper and copper glance" or sulphuret. For the most part the geological formation of New Mexico assures wealth in the form of a number of minerals, and not the least is that of copper.

Arizona.—This Territory has also mines of copper, which are found throughout her domain. A large deposit occurs in the valley of the Gila, north of Tucson, not far from the line of the Southern Pacific Railway. These copper-ores are in great abundance, and are of almost every class, such as sulphurets, carbonates, and oxides; some of them carry a small quantity of gold and silver. As a rule, the copper-ores of Arizona occur in diversified forms; some are found in isolated chambers or pockets, as the miners say, while others are in well-defined veins, and may be traced by their regularity for a long distance. There are also numbers of localities where indications of copper occur. The time may not be far distant when this great mineral wealth will be utilized for the benefit of the people.

Copper on the Pacific Slope.— In the Puget Sound region, in Washington Territory, occurs a large vein of copper-ore carrying silver, and, ac-

15

cording to the Governor's report (1886), that ore abounds in several portions of the Territory; but definite information on the subject of the amount and quality had not been ascertained, as these mines had been only partially opened and worked. In due time these copper deposits will be made available and utilized, but at present the people are much more absorbed in mining coal and iron-ores than in any other class of minerals.

In Oregon deposits of rich copper-ores exist in various places; these have been occasionally opened and but partially worked—the people having been much more engaged in other pursuits, and what mining they have done has been for gold in the eastern portion of the State, which for the present pays much more than mining copper, even if the facilities for it were much better than they are. Copper can be obtained so easily from other portions of the Union, that on the Pacific slope its mining, commercially speaking, is not satisfactory.

In Nevada County, California, copper deposits of great value are located. The ore is a sulphuret and for the most part lies in an inaccessible position for mining. There are several other localities in the State where copper is found, yet the whole State mined only about 235 short tons in 1885.

XXI.

THE galena or ore of this metal is abundantly distributed over the United States, but very unequally as to the quantity found in different places. It sometimes occurs between strata of limestone, but oftener in veins or deposits in connection with other metals, especially silver and copper. Though it is found in numerous places on the Atlantic slope, the amount is limited, our main deposit being within the valley of the Mississippi and in the Rocky Mountains. Geologically speaking, it occurs among several kinds of rocks and of different ages and formations, and appears to be at home in any of them; but galena, or the sulphuret of lead, is specially its own ore, since it is never obtained pure, like copper, or silver, or gold.

Lead on the Atlantic Slope.—In New England lead is found in every State except Rhode Island. In these States it has been mined in comparatively small quantities and at heavy expense, and only to supply a domestic want. In working these lead-ores a small amount of silver has sometimes been obtained. Much richer mines than those of New England have been discovered and worked in Northern New York, in St. Lawrence County. In the latter deposit the galena is famed for its beautiful crystallizations, but the production of the metal in respect to profit has not been successful. Numerous other small veins of lead-ore are also found in this State. The deposits of lead east of the Alleghanies are often deceptive. They are frequently isolated in what the miners call " bunches "; if a vein happens to be continuous, it is more likely to become thin than otherwise, and not productive enough to warrant the expense of removing the surroundings of the ore. At one time lead-mines located in Chester County, Pennsylvania, were thought to be valuable, but they developed the usual lack of ore, and finally were abandoned. In the gold region of North Carolina one lead-mine was discovered and worked a greater length of time than any other on the Atlantic slope, but finally it became exhausted. The ore in this instance was combined to a small extent with silver.

Galena is found in the middle and northeast portion of the State of Georgia, though in localities widely separated, but in all more or less sparingly. The ore when pure contains 86.6 per cent of lead and 13.4 of sulphur. In one mine it is found associated with silver, gold, and copper; and in another with gold alone, and often with silver and copper; then, again, in small quantities amid ledges of limestone as well as within layers of sandstone. The lead-ore of this State nearly always contains some portion of silver, and sometimes to such an extent as to render it quite valuable as a silver-ore, though the quantity is quite limited.

The Western Lead-Fields.—The prominent deposits of lead-ore in the Great Valley are limited to two—one on the upper Mississippi, and the other 300 miles directly south in the States of Missouri and Kansas. These deposits, though possessing some qualities in common, are as geologically distinct as they are far separated. The first field includes adjacent portions of three States—Wisconsin, Illinois, and Iowa; the second comprises mines in Missouri, principally on the branches of the Maramec River, southwest of the city of St. Louis; and still farther to the southwest, where the lead-ore is combined with that of zinc. The lead region of the upper Mississippi is 4,800 square miles in extent. What is termed the surface ore abounds throughout the region, and oftentimes plowmen turn up lumps of it in the fields. " There are no deposits of lead in this region that come under the head of true veins; they are invariably limited in depth, and they are all in a certain geological formation, and the productive part does not generally exceed one hundred feet in thickness." (*Whitney's Metals, etc., p. 410.*)

Lead-Mining.—The surface ore becoming praetically exhausted, the lead was sought in the native deposits, and its mining commenced on a large scale. The cities of Galena and Dubuque—the one in Illinois and the other almost opposite in Iowa—are both much interested in this mining industry. The smelting of lead is proverbially an unhealthy business. The furnaces are placed, if possible, remote from human habitations, and the chimneys are very high, in order to carry off the poisonous fumes. In the vicinity of Galena the chimneys extend sometimes to the top of the hill from furnaces in the valley below; the fumes from these will even wither the grass, and the only animal that is proof against lead-poison is the swine. Great numbers of the miners at first were foreigners, originally for the most part from mines in the north of England and from Cornwall. These deposits run deep, but the ore is followed down by the persevering miner with his pick and sledge and drills, for he often calls to his aid blasts of gunpowder to loosen the masses of ore.

The Mound-Builders and Lead.—As no *pure* lead, but only galena, or lead-ore, has been found in the mounds located in the vicinity of these mines, it has been inferred that their builders knew of the metal; the inference is rather that they knew only of the ore, and were ignorant of the art of smelting it. It is much more probable that the lumps of galena were found on the surface or near it, and placed in position in the mounds, as the builders were in the habit of putting stones in these burial-places.

Copper has also been found in the mounds; but that metal the builders could have obtained in a pure state, but not the lead. There is only one instance in which it has been supposed that a lead-mine—the Buck Lode—near Galena, had been worked before the coming of white men.

Missouri and Kansas Lead.—The lead area— that is, where that metal is found—of the State of Missouri is immense. It extends in two belts: one in a southerly direction from within a score of miles of St. Louis, in that portion of the State which lies between the Mississippi River and the Ozark Mountains, and along the latter's eastern slope down nearly to the State boundary. This region southeast of Pilot Knob, except near the river, is rough and hilly, but abounding in isolated deposits of minerals, such as lead, iron-ore, brown hematite, copper, nickel, and zinc. The other belt extends in a southwest direction through the middle portion of the State, on the head-streams of the Maramec and Osage Rivers, and thence along the western slope of the Ozark Mountains to the extreme southwest portion of the State, even into Southeastern Kansas, into Arkansas, and also into the Indian Territory. This rich southwest deposit of lead-ore is also combined with zinc—the latter predominating, as we shall see.

In 1875 lead-ore was accidentally discovered in Dade County, in the southwestern portion of the State. It was an immense and solid mass of lead almost pure, and amounting in weight to about twenty-five tons; this discovery was as surprising as that of the great mass of anthracite coal on Mauch

Chunk Mountain in Pennsylvania (page 30). The ore in this region is usually found in smaller quantities, and in isolated deposits at depths of a few inches to one hundred feet, and in this particular locality occupying an area of two square miles, with indications of similar large deposits in the vicinity. The quality of the lead and zinc ores thus associated is deemed the richest within the State. There are also in this section many known deposits of lead and zinc ores waiting in store for future use, as their contents are not yet needed to supply the wants of the people; when that necessity arises, railways will make them available. The cities of Joplin and Carthage, in Jasper County, are both prominent as being each the center of a rich mineral district. The village of Granby, in Newton County—among the first to enter upon the work—is largely engaged in mining lead. The ore lies unusually deep, and shafts are sunk in order to reach it, but its great richness amply repays the extra expenditure. Adjacent to these lead-mines are those in Cherokee County, Kansas, in which zinc-ore appears to predominate; hence in the latter the output of zinc is greater than that of lead. In speaking of Central Missouri, including the western slope of the Ozark Mountains, the " Handbook " of the State says: " Lead [combined with zinc], iron, and copper ores crop out of all the hills and bluffs as well as showing on the surface in the rich valleys; .. . the lead and iron ores are in inexhaustible quantities."

New Mexico and Arizona Lead.—In New Mexico, in the vicinity of Socorro, in the valley of the Rio Grande, is an extensive deposit of lead-ore which has been opened to some extent. This ore is rich and combined slightly with silver, while the expenses, for the present, of mining and smelting, are comparatively small. Lead-ore, apparently of the same general character, is found also in Arizona, and quite extensive. These mines have been operated to some extent.

Rocky Mountain Lead.—The lead-ores in the States of Illinois, Wisconsin, Iowa, Missouri, and Kansas are more in combination with zinc than with any other metal, but in the Rocky Mountains they are closely allied with silver, and to such an extent that mineralogists distinguish the lead-ore of the Rocky Mountains as argentiferous or silverized, while that from the States just mentioned is characterized as non-argentiferous or devoid of silver. The distance from Joplin, near the center of the Missouri lead region, to Leadville, near the center of that of the Rocky Mountains, is about 700 miles, while the two fields in their direction north and south are nearly parallel. The amount of lead derived from the mines in the Rocky Mountains, in 1884, was about six times as much as that from the mines in the valley of the Mississippi, while those of the latter not only produced lead, but compensated the miners by their output of zinc. As yet scarcely any zinc-ores have been discovered in the Rocky Mountains, while an authority states that the silver produced from the lead-mines pays every expense, so that the lead obtained is clear profit.

Colorado and Utah Lead.—Colorado produces more lead than any other State in the Union, and Leadville is the present center of that production. This furnishes also both silver and gold, as these are here allied with the lead-ore. In 1884, 63,165 tons of lead were mined in this State—the Leadville district alone furnishing 35,296 tons, with 5,720,904 ounces of silver, and 22,626 ounces of gold. There are certain classes of ores found in the vicinity of Leadville wherein exist other elements, such as zinc-ore to a small extent, and sulphurets of iron as well as of silver. New discoveries continue to be made in the region and new mines opened, the lead area being very extensive. Utah has also a large area where lead-ores are obtained; they partake of the usual qualities of the Rocky Mountain ores, though they are located within the Great Basin, and on the west slope of the latter mountains. A large number of these deposits have been opened, and the ores mined to such an extent as to make the Territory a large producer of lead. The ores, however, of Utah are not of as high grade as those of some of her neighbors. It is often the case that the value of the lead obtained from these mines is less than that of the silver.

Montana and Idaho Lead.—Lead is found in South-Middle Montana near Helena, 500 miles west of

north from Leadville. This ore is in veins or fissures in the granite, and ranges in thickness from
one foot to one hundred. This galena or lead-ore is
associated to some extent with that of zinc, as well
as with copper and the sulphuret of iron, silver being
very rare. Lead deposits also exist in the northern
portion of Montana, on the head-streams of the Missouri. In Idaho the deposits are amid ledges of
limestone, and being very irregular the former do
not conform to the layers or dip of the latter. Thus
we see the remarkable variety that prevails in the
forms and associations of the lead-ores found in this
region, which, though of unequal width, extends
from about the thirty-third parallel of north latitude,
in New Mexico and Arizona, across Utah and Colorado, to the forty-seventh parallel in Montana and
Idaho—stretching north and south amid the Rocky
Mountains and along their western slopes for nearly
1,000 miles. The deposits of lead-ores in this immense area are by no means connected with one another, as there are often intervening spaces in which
lead has not thus far been discovered. The presumption is that hereafter many more such deposits will be brought to light, and their treasures
utilized to supply the wants of the people. The
entire area within the United States where lead-ore
is found is very difficult to be ascertained correctly, but it is thought that 50,000 square miles is a
fair estimate. Be that as it may, "the United
States now occupies the first rank in the world in

the production of l'ead." (*Mining Journal, vol. xxxvii, p. 289.*)

Demands for Lead.—In the advancing progress of the United States, lead, in numerous forms, be·comes more and more necessary for the comfort of the people. In villages and cities that are supplied with water by means of pipes leading from reser-voirs, the amount used is enormous; and as villages expand into cities and grow in population, is the demand increased, as they all are anxious to be thus supplied with water. This demand leads sometimes to undue stimulation in mining, and a discrepancy occurs in the amount of output in different succeed-ing years, and, instead of a uniform increase, there may be a falling off in the production, and then again an unusual advance. This does not result from a deficiency in the amount of available ore, generally, but from the prudence of the miners, who find that there has been an overproduction, and they curtail their output accordingly.

XXII.

ZINC.

THE ores of zinc are always found in combination with those of other metals, such as lead, which is often blended with silver, or with copper pyrites. The zinc-ores do not generally contain as much silver as they do galena or lead-ore, and usually the zinc-ore, as the vein runs deep, appears to gain on the lead. In the State of New Hampshire zinc seems to be distributed wherever there are ores of other metals, but in quite limited quantities. In time science may enable the miner to utilize these ores.

In New York State an ore deposit of blende or sulphuret of zinc is found in Sullivan County. The metal-bearing contents of the vein are sulphur combined with zinc, lead, copper, and iron, with traces of silver; these are associated with crystallized quartz, the zinc and lead predominating: in this instance the lead of solid ore varies from a mere seam to three feet in thickness. There are, also, many other places in the State where zinc-ore occurs, but in small quantities; as in lead-mines in St.

Lawrence County, and in the vicinity of Niagara Falls, where it is found in small crystals, while at the famous Stirling Mountain, Orange County, it occurs in connection with iron-ore.

The Great Zinc Deposits.—The most extensive mines of zinc on the Atlantic slope are located in Sussex County, New Jersey. These deposits extend along a range of hills in a southwest direction from near Stirling Mountain in the State of New York into the State of New Jersey, to Franklin, where is produced the unusual metal known as Franklinite—a compound of iron, zinc, and manganese, the latter being a peculiar metal of a whitish-gray color, very hard and difficult to fuse or melt. Here is one of the finest zinc and manganese ore-beds yet discovered in the Union, if not in the world, and here are produced manganese, iron, zinc, or spelter, and the oxide of zinc. The latter is used extensively in the manufacture of a brilliantly white paint, the effect of which is not only striking as used in ordinary painting, but in giving the purest white tints when employed in making wall-paper. When prepared for use it is "a clear, white, impalpable powder," and it is claimed to be chemically pure. Metallic zinc is also produced in the form of slabs and plates, as it is susceptible of being rolled into sheets.

The ore-beds at these mines vary often in thickness, ranging during a year's workings sometimes from two to thirty feet. Great improvements have been made in the methods of extracting the zinc from its surroundings and combinations with other metals, and the latter with sulphur.

Pennsylvania Zinc.—The Jersey hills in which zinc-ores occur extend into the State of Pennsylvania, and nearly the same class of ores are found in Lehigh County, though their characteristics are somewhat different. In Pennsylvania the ore is a silicate of zinc of a good quality, and quite free from mixture with either lead or iron. Here the deposits are immense; the ore lies in the same direction as the adjacent limestone. Smelting-furnaces are near the town of Bethlehem. These more especially make the oxide for painting directly from the ore, and the latter yields from 40 to 60 per cent of metal.

Tennessee Zinc.—Zinc-ore occurs in different places amid the Cumberland Mountains in the eastern portion of this State. The deposits are found to run for some sixty or seventy miles in a southwest and northeast direction, extending even into the State of Virginia, where the ore is said to be of good quality. The deposits are within the valleys of the Clinch and the Holston Rivers; these two unite and form the Tennessee. Works have been established to manufacture metallic zinc, though the ores are said not to be as rich as some others, owing to the presence of sulphur. One of these establishments has a capacity to turn out a ton and a half of metallic zinc a day.

Western Zinc.—We now pass to the valley of the Mississippi, and find zinc-ore deposits of great value and extent in the north-middle portion of the State of Illinois, in the vicinity of the towns of Peru and La Salle; in the State of Wisconsin, and southwest of these beyond the Mississippi within the States of Missouri, Kansas, and Arkansas. In all these the ore is more or less combined with that of lead.

On the western slope of the Ozark Mountains, in Missouri, is an extensive area where zinc-ore abounds. These beds are in the Southwestern portion of the State, and extend into Southeastern Kansas, and probably still farther into the Indian Territory. The zinc deposits in Dade County are stated in the "Handbook of Missouri" to be "practically inexhaustible."

Discoveries were made, in 1873, of zinc-ore in Southwestern Missouri, in Jasper County. Here are located mines or deposits of both lead and zinc ores combined, and in immense quantities; they extend into the adjoining county of Newton and across the State line far into Kansas. These lead and zinc ores are the richest thus far discovered within the United States. The city of Joplin is the center of this important mineral district, which for the last few years has furnished three fourths of all the zinc manufactured in the United States. The entire region is heavily timbered, and well watered by living streams, in and along the valleys of which are the mineral deposits. The city of Joplin, because of its extensive operations in mining zinc and lead, has become the fourth city in size in the State of

Missouri. The output of these mines, in 1884, was 74,250 tons of zinc-ore, which was transported to smelting establishments, while in the adjacent county of Cherokee, in Kansas, it is estimated that there were mined about 33,000 tons of the same ore. Numerous discoveries of zinc and lead ores have been made in other places in Central Missouri and also in the counties of the State of Arkansas that border on the Missouri line. The United States ranks third among the zinc-producers of the world.

16

XXIII.

THIS metal, when in a state of ore, combines with very few substances; only two of such combinations are worthy of mention in this limited notice—one, *tin-stone;* the other, *tin pyrites*, or sulphuret of tin. The former is an oxide, and in a hundred parts contains 78.62 of tin and 21.38 of oxygen; the latter is usually allied with copper, or iron, or zinc, and often at the same time with all three. The ore when imbedded in quartz-rock and mica or tin-stone, or in the fissures of a certain class of granitic rocks, is, in color, usually dark brown or even black. The metal itself is a silvery white, with a yellow tint; it is harder than lead, but softer than zinc. It can not to any extent be drawn out into wire, but can be beaten into leaves so thin that 1,000 of them could be comprised within the thickness of an inch. When a bar of tin is bent, it produces what is called the " cry of tin," and experts can judge of its quality by this sound. " The purest form of tin in commerce is called " grain-tin," and the ordinary quality " block-tin." The geological position of tin-ore is peculiar, as it is found only in the three oldest class of rocks.

Tin ; when discovered.—Tin was known from the earliest historical times, and combining it with copper was practiced by the ancients; as we find bronze was in use in the earliest historical periods, before they made brass by combining copper with zinc. The most important district in Europe for the production of tin is Cornwall, in England. The mines in that vicinity were known and worked by the Phœnicians before the Christian era, and have been worked from that day to this ; though now the excavations extend for about 3,000 feet into the earth, and leads are driven in every direction, some far out under the ocean. In modern times an immense amount of tin is obtained from islands in the Malayan Archipelago, the principal deposit being on the small Island of Banca. These mines are worked almost exclusively by the Chinese, who exercise but little engineering skill, as the excavations are shallow and worked from the top only, the ore being what is termed stream or alluvial ore.

Tin in the United States.—Until recently, tin has never been discovered in the Union, but in small quantities and in isolated places—a few crystals being found here and there amid granitic rocks, but of no commercial value. In the vicinity of Jackson, New Hampshire, tin-ore exists in veins ranging in thickness from one inch to eight. It has been found also near Windsor, in the State of Maine; at Goshen, in Massachusetts; and at Haddam, in Connecticut; and in many other places, traces of tin are found. At Windsor attempts were made to work the mines, but without success, because of the limited amount of ore. In the magnetic iron-ores of New York and New Jersey, tin in small quantities has been detected by chemists.

Tin in Virginia.—In Rockbridge County in this State, in one place, a "tin-bearing vein apparently cuts across the rock for thirty-six feet. This vein is made up of white quartz with tin-ore, but the thickness is not stated." The samples assayed showed an average of 32 per cent of tin. In the adjoining county of Nelson "is a vein several inches thick running east and west, and which traverses a rock consisting principally of large crystals of quartz and feldspar with some mica." On the northwestern slope of the Blue Ridge is a tin deposit or vein in granitic rock at an elevation of 2,700 feet above the sea. This mine is twenty-two miles east of Lexington, the capital of Rockbridge County. "The main lode extends in the general direction of the mountains and rock formations. It is apparently made up of several parallel veins with an average width together of perhaps 100 feet, and traceable on the surface." The workable ore, so far as shown, exists practically in lumps and veins clearly definable, and the color of the ore, which is free from injurious minerals, is a light yellowish brown to a dark brown. "Selected pieces of this tin-stone yielded more than 70 per cent of metallic tin." In several places operations have been recently commenced to extract the tin; but the veins, though rich, are so thin, as a general rule, that fears are entertained they will not be remunerative. (*Mineral Resources of U. S., pp. 599–602, for 1883–1884.*) Within recent years tin-bearing veins have been found in that section of the country in a number of places, especially in Mason and in Cabell Counties, West Virginia. Tin-ore has been discovered near King's Mountain, North Carolina, and under the usual conditions; also minute particles in Georgia and Alabama. These discoveries have all been made within recent years, and future exploration may lead to finding large deposits of tin-ore. This metal is said to have been found within a few years, but in small quantities, in the States of Texas and Missouri.

Tin in the Black Hills.—The people of the United States, though abundantly supplied, as we have seen, from their own resources, with the useful minerals of coal and iron, gold and silver, and copper and lead, yet of tin they virtually knew of none within their wide domain. But the influence of a discovery in 1883 may make them as independent of the outside world in respect to tin as they are in relation to the other metals. In June of that year there arrived at San Francisco a box filled with specimens of a heavy, dark-colored ore; it was sent from the Black Hills of Dakota, and was consigned to a firm in that city, with the request to assay the ore and determine its character. It was tested, and found to be *cassiterite*, or tin-ore, and that it assayed 40 per cent in metallic tin. The location of the discovery is in a central position amid the Black Hills— thus named from the peculiarly dark foliage of the dense growth of pines with which they are covered. The meridian line of 104° west is the boundary between Dakota and Wyoming Territories; this line gives about two thirds of the Black Hills to the former.

They stand amid immense plains; and the internal convulsion that pushed them up also brought to the surface at their tops some of the oldest rocks, geologically speaking. These rocks rise in groups of irregular peaks and broken ridges, running north and south in their general direction. They consist more or less of fine-grained mica-schist, a slaty structure, micaceous sandstones, while traversed by veins of quartz slightly gold-bearing. Amid these various strata is found tin-stone, in which are seen little globules of tin-ore, generally of a dark-brown or even almost black color. The entire geological formation is very much diversified. But we can not go into detail.

Numerous discoveries of tin-ore have been made in the Hills, as the tin area covers many square miles of surface; this area is in progress of being thoroughly and scientifically explored. From lack of the proper knowledge, many have located claims that have proved worthless. The Black Hills are by no means one vast deposit of tin-ore. The portion of the Hills in Dakota has been much better explored or prospected in respect to tin-mines than the portion belonging to Wyoming. Hon. Francis E. Warreu, Governor of the latter, in his report (1885) to the Secretary of the Interior, affirms that tin-ore exists equally on the west slope of the Hills, but the deposits have been thus far developed only sufficiently to determine their general character and scope.

Placer-Tin.—As in finding gold there are placer-mines, so are they in finding tin, the difference being that the gold obtained is pure, while the tin is in the form of ore. The latter is called "stream-tin," because found in the beds of streams, or in the alluvial soil along their banks. In ages past the elements disintegrated the peculiar rocks and veins in which the tin-ore was imbedded, and, as the smaller pieces were carried down the mountain, the attritions caused by the rushing waters stripped the ore more or less of its surroundings, and being thus free and heavy it sank to the bottom. The usual placer tin-ore is found in globules of different sizes, from "that of a pea to that of a pigeon-egg, and even sometimes larger." The miners for gold in some of the placers of the Black Hills were often inconvenienced by the presence of larger masses of this heavy black ore, or as they thought stone, and, not knowing its value, they threw it out of the way as a hindrance. "It is now probable that in these placers both the stream-tin and the gold will be mined together and to much greater advantage than either worked alone." (*Min. Res. of U. S. for 1883–1884, p. 613.*) In the Black Hills, on both sides, tin placer-mines are found along the streams that have their head-springs amid the "greisen rocks carrying tin-stone."

The Etta Mine.—Among the many locations of tin-mines on the eastern slope of the Black Hills, that of the Etta is now the most extensively worked, and may be deemed a successful pioneer in this new industry, and for that reason deserves a brief notice. This mine is in the central portion of the Hills, in Pennington County, Dakota, a few miles east of Harney Peak and about twenty southwest of Rapid City, and also near the dividing range of the region. The surface opening is on an isolated granitic hill, which rises like a cone for 250 feet above the surrounding valley, though it is 4,500 feet above the ocean. This whole region is remarkable for its picturesque beauty of hills and valleys and numerous streams of living water.

A granitic rock known to geologists as "greisen" carries with it "tin-stone," and together with it occurs mica; in the midst of these compounds is imbedded tin-ore in irregularly shaped lumps, varying in size from a grain of wheat to a quarter of an inch in diameter, and sometimes larger. "Massive tin-ore was found near the top of the hill while excavating for mica. Some of the heavier pieces of the ore obtained weighed from fifty to sixty pounds, while lumps were common whose weight was only three or four." The veins of the "greisen tin-ore" are here nearly vertical, running downward into the mountain. They are reached by tunnels from the side. One of these entering at a distance of fifty feet from the hill-top struck a body of greisen tin-ore over twenty feet thick; another entering at a point 125 feet still lower down also reached the same vein, which proves that it runs downward similar to the veins in the mines in Cornwall, England. This characteristic appears to apply to all the tin-ore veins that have been investigated in the Hills. Thus far the vein-mining is quite easy, because of these horizontal openings; but the time will no doubt come when the tin-ore of the placer deposits will be exhausted, as well as that obtained by means of side-tunnels; then the miners will go below the surface, as they do in Cornwall.

The **Tin extracted; how?**—Tin-bearing rock is subjected to a stamping process similar to that applied to gold-bearing quartz. The mass of crushed tin-ore is passed through washings in which the pulverized mica, spar, etc., are carried away, while the heavier tin-ore sinks, and is caught in pockets designed for the purpose. The heavy crystals of tin-ore separate easily from their surroundings of mica and other ingredients, and the pure ore thus obtained is known as "black tin," which is now ready for smelting, the result being metallic tin of a bright or silvery color.

The ore of these mines is remarkably free from

any minerals or impurities, such as sulphur or arsenic, which tend to impair the quality of the tin; neither has it any mixture with the sulphurets of iron, or of lead, or of copper. These general characteristics, as far as known, appear to belong to the ore of the tin area on both sides of the Black Hills. In these several respects just mentioned this tin-ore is found to be cleaner and purer than that derived from the mines of Cornwall, while the percentage of metallic tin obtained from the pure ore, or black tin, is about the same in both cases—that is, ranging from 60 to 70 per cent. Numerous tin deposits have been discovered along the eastern slope of the Black Hills, and in many instances mining operations have been commenced. The area wherein these tin mines are found is estimated at more than 100 square miles.

Tin in Wyoming Territory.—On the west slope of the Black Hills tin-ore appears to be abundant, while as far as tested it is found equal in quality to that mined on the east side. "There are no heavy impurities in the veins with which to contend; no combinations of tin with sulphur or arsenic. . . . The granite intrusions are more regular and lode-like than on the eastern side of the range, and the crystallization is not so coarse." (*Min. Res. of U. S., 1883–1884, p. 611.*) Mines have been opened near Hill City, in the extreme northwestern portion of the Harney Peak tin region, stream-tin of a high grade having been previously picked up along and in the creeks running northwestward to branches of the Missouri. This ore when assayed was found to yield about 46 per cent of metallic tin. The area of this particular tin district is about thirty square miles.

Idaho and Montana Tin.—In many points in these Territories tin-ore has been found, and the indications are that it exists to a great extent within the region of the Northwest, where granitic formations prevail, particularly amid the Rocky Mountains, on the head-waters of the Missouri, and also west of them on the streams flowing from the Bitter Root Mountains into Snake River. " A recent discovery of 'float' masses of tin-ore, assaying 60 per cent, is reported from Cœur d'Alene, Idaho." Tin-ore has also been found in Montana, near Helena, the capital, and likewise in other places; but these deposits thus far have not been developed.

In Southern California tin-ore has been found in San Bernardino County, southeast of Los Angeles. It is a region of granitic rocks, and tin-stone occurs in many places, but "in small quantities and in various veinlets and small lodes. . . . This region, though affording tin oxide in many places over a considerable area, has not realized the hopeful expectations of those who have made efforts to work such mines." A number of places in Middle and Northern California have been named as having deposits of tin-ore, but nothing definite or reliable has yet been ascertained on the subject.

XXIV. ·

Chromium, or Chrome.—This is defined as a hard, brittle metal, of a grayish color, and related to iron in many of its properties. It derives its name from a Greek word meaning color, and its compounds produce various and beautiful colors, as in the fine, deep green of an enamel on porcelain and in glass, while in the chemistry of Nature it is traced in the green color of the emerald, as well as in many instances in the arts. It is used for decorative purposes.

About the year 1825 Mr. Isaac Tyson discovered on his farm, in Baltimore County, Maryland, a peculiar substance which proved to be "chrome iron-stone" or ore. This ore was at first mined and sent to England, which seems to have been the only place where it was in demand. About twenty years after this discovery of the mine, the Tysons, father and son, commenced the business of decomposing the ore, and obtaining from it various pigments or paints; hence we have chrome - yellow, chrome - orange, chrome-green, and even Turkey-red, and other colors used in printing calico. The family (in the fourth generation) still continue (1886) that business in the vicinity of Baltimore. In making experiments they acquired several important secrets (that may be termed family), in respect to manipulating the ore, that enables them to produce the pigments at less expense than can others, and now they alone virtually supply the American market.

Several other mines of chromium-ore were discovered in the State; these the Tysons secured, if after investigation they proved valuable. The deposits are all liable to be soon exhausted, though one of them, in Harford County, furnished ore steadily at the annual rate of 2,500 tons for more than forty years. Mines were also discovered in Lancaster, Delaware, and Chester Counties, Pennsylvania. Chrome-ore is noted for occurring in widely separated localities, though not in leads or veins from the direction of which others might be discovered, but in "pockets" of different sizes, ranging from a few pounds to thousands of tons. Small isolated deposits occur in the States of Massachusetts, Vermont, and New York, and south of Maryland in Virginia, and in Jackson County, North Carolina. The latter is reported exceptionally rich in chromium.

On the Pacific slope large deposits of chrome-ore have been found in Del Norte, Placer, and Sonoma and other counties in California, and these now supply ore to the Tyson manufactory at Baltimore, to such an extent that the working of the mines on the Atlantic slope has for the present been practically abandoned. This ore comes from San Francisco in sailing-vessels round Cape Horn. So large a number of chrome deposits, great and small, occur in California, that "chromium is now known to be common throughout the State." (*Min. Res. of U. S., 1883–1884.*)

It may be stated that what is known as chrome-steel is made from the iron-ore that is associated with chromium in the mine, and that the latter imparts to the steel certain properties, though much the greater portion of the chrome-ore is utilized in preparing pigments.

Platinum.—Platinum is found in small quantities in the United States, usually in pure grains and frequently with iridium, and is associated with gold in placer-washings. The color is that of silver, but less bright. Its use is quite limited, because it is made available in only small amounts, as in pins for artificial teeth or in tips for lightning-rods, and in certain cases in electric lights, and in giving luster in porcelain-painting. The great heat required to melt platinum, and its property of resisting the action of acid, make it essential that numbers of the utensils in the chemist's laboratory, such as crucibles, retorts, etc., should be made of that metal; hence much of its consumption consists in furnishing chemical apparatus. As platinum has but little affinity for other substances, it is rarely found in the form of ore, but in this country in small pure grains mixed with gold sands, and therefore the product is much limited in the United States, that of 1885 being about twenty pounds troy. The American ores appear superior in fineness to those obtained from the Russian mines in the Ural Mountains—the former giving 85.5 per cent of platinum, and the latter 76.4. Russian mines, however, produce so much greater quantity, that they furnish about 80 per cent of the world's production, and yet the general consumption of the metal is so limited that this amount is an overproduction.

The particles of platinum, following no rule, are scattered in bewildering confusion in foreign lands as well as in the United States. On the Atlantic slope a nugget, weighing 4.35 pennyweights, was picked up near Plattsburg, New York, while minute grains of it have been found associated with gold in Virginia, North Carolina, Georgia, and in other Southern States, but it occurs in much larger quantities on the Pacific slope, especially in the States of California and Oregon and in the Territory of Idaho. In the first State it has occasionally been found in unusually large lumps, weighing between *two and three ounces.* In some instances in the hydraulic mines, for *nine* ounces of gold *one* of platinum has been obtained in small grains. " On the Oregon coast the proportion of gold to platinum in the placers is sometimes *five* to *one*, and in rare instances the amount of platinum equals the gold." (*Min. Res. of U. S., 1883–1884, p. 576.*) *Iridiomine* is found in connection with platinum, and is used for pointing gold pens.

Iridium. — In 1885 twenty-five pounds troy of this metal was mined in the United States, principally in Oregon. Iridium is found in intimate relations with platinum, of which it is sometimes an alloy, but more frequently with the latter when it accompanies gold in placer-washings. It has a whitish color and luster somewhat like steel, and is also very hard and one of the heaviest metals. It has no affinity for acids except when alloyed with osmium or platinum, and to melt or fuse it requires greater heat than platinum. It is used in making what are called "diamond points" on gold pens and on stylographic pencils, and for a similar purpose it is applied to instruments used by surveyors and engineers ; also "the electrical points of the telegraphic apparatus" are made of iridium. To avoid rust or corrosion, hypodermic needles for physicians are now made of gold and tipped with iridium. It is employed in making portions of various instruments where delicate operations are required, such as the tiniest scales of the druggist, and under certain conditions it makes a black pigment used for decorating fine porcelain, and sometimes when alloyed with platinum it is used for chemical purposes.

Russia, thus far, is the chief producer of iridium. In the placer-mines of the Ural Mountains it is found associated with gold and platinum, and in precisely the same relations it occurs in the United States. Though in very limited amounts, iridium occurs in the East Indies, in Australia, in South America, and in Europe; in addition to Russia, small quantities are found in Germany, France, and Spain, and also in Canada. In our country the principal producers of iridium are the States of California and Oregon. The indications are that if the demand for this metal would warrant the effort, the miners in California would "prospect" for it; but for the most part they are ignorant both of the properties and of the value of the mineral, and in consequence it is thrown away as refuse. On the authority of the State mineralist, Mr. Hanks, it is asserted that it is found amid the river-sands of the northern counties of the State. "Considerable quantities accumulate ın the mints and assay-offices obtained from the crucibles in melting placer gold." (*Min. Res. of U. S., 1883–1884, p. 582.*)

Nickel.—The two metals—nickel and cobalt—are intimately connected. Nickel is a grayish-white metal; it is extensively distributed over the world, and is also found in meteoric stones; it exists in the atmosphere in what scientists call "meteoric dust," while "its presence in the sun is revealed to us by the spectroscope. . . . It probably pervades the solar system." Nickel has considerable luster, so that "when pure it is susceptible of a brilliant polish that will remain untarnished for a long time." When pure it has been manufactured into hollow-ware as culinary vessels. And experiments made during this century prove that it is adapted to numerous uses. Among these is the important one of its alloy

17

with copper and zinc, known as nickel or German silver, first introduced in Berlin, as a substitute for silver because less expensive. This strong and white alloy is manufactured into table furniture to take the place of the usual silver-ware, and into forks and spoons, etc. The art of nickel-plating by the galvanic process was likewise introduced; the basis being the nickel-silver which takes and holds the pure silver itself. It is also used in the manufacture of glass, giving to it a green color.

A nickel alloy is used for money or coin in Belgium and in the German Empire. The United States also use, for the subsidiary coins, nickel alloyed with copper, first in the cent, which contained 12 parts of nickel to 88 of copper. These were withdrawn or called in in 1865, and instead were issued *three* and *five* cent pieces, containing each 25 parts of nickel to 75 of copper. The *three*-cent pieces, however, are no longer coined.

Cobalt.—Cobalt is a metal of a reddish-gray color, brittle and hard to melt; it is combined with other substances, as sulphur, arsenic, or iron, but principally with nickel. No American ore is worked for cobalt alone, the latter being secondary in importance to whatever ore it is associated with, be it nickel, iron, or copper. The uses of cobalt and its compounds are quite limited; under certain conditions, as the oxide, it forms blue colors, and glass-manufacturers use it in giving a blue tinge to that article, and potters to correct the yellow tint incident to that ware, as well as in producing blue ware or for ornamentation. Cobalt oxide is much used for all kinds of decorative work on pottery, and a certain compound containing cobalt produces a green color.

Nickel and Cobalt Deposits.—These ores in combination occur in detached localities and in limited quantities on the Atlantic slope from New England to North Carolina. More than one hundred years ago a deposit of nickel and cobalt ores was discovered near Middletown, Connecticut. This mine was worked specially to obtain cobalt; but in 1820, when 1,000 pounds of the product, then supposed to be cobalt, were sent to England to be refined, it proved to be nickel, in connection with very little cobalt. Small quantities of nickel-ore have been discovered in two other places in the State—Torrington and Litchfield; both these mines have been long since closed. In New York State traces of cobalt occur, in connection with iron pyrites (sulphuret), at Anthony's Nose on the Hudson River.

The most important deposit of these ores thus far found east of the Alleghanies is at Lancaster Gap, Lancaster County, Pennsylvania. The ore of this mine, after certain preparations, is shipped to Camden, New Jersey, at which place are the most extensive and complete appliances for extracting both the nickel and the cobalt from this mixed ore. In the refining process, as conducted here, nickel was obtained in a pure state (1872)—a feat of chemistry never before attained. We have already mentioned the uses to which pure nickel is applied.

Nickel is found also in Lehigh County, Pennsylvania, and in Maryland, though in small quantities, and in connection with chrome-ores, while there has been discovered little more than a trace in North Carolina. In Madison County, Missouri, and at the mine La Motte, cobalt is found in connection with its usually accompanying minerals; but also, with the black and the brown oxides of manganese, at a certain depth of the mine, nickel appears. The combination of nickel and cobalt ores occurs in New Mexico, and no doubt will yet be found in Arizona, in Colorado, Nevada, California, and Oregon. The mode in which these ores sometimes appear is singular. For illustration, in Gunnison County, Colorado, is a deposit of cobalt with scarcely a trace of nickel, but in connection with a number of other kinds of ore, while the cobalt itself is only about 12 per cent of the whole; and again, near Silver City, in the same State, is a deposit of nickel-ore, mixed with little more than a trace of cobalt. The State of Nevada has a number of deposits, though the proportions in those of the nickel and cobalt are by no means uniform. Nickel-ore has likewise been discovered in a belt of country in California extending for twenty-five miles, and lying east of the higher granitic mountains of the Sierra Nevada near Lake Mono. This ore has 34 per cent of nickel. These combined ores also occur in a few places in Southern California. In Douglas County, Oregon, are deposits believed to be large, though as usual scattered over quite an extensive area.

Antimony.—This metal is of a tin-white color; it is quite brittle, but hard, and can be reduced to a powder. It rarely occurs in a metallic or native state, but in combination with different substances, oftener with sulphur than with others. Its ores have various colors, as gray, white, or red, and their many shades. Because of its volatility, and great affinity for oxygen, it is difficult to obtain the metal pure from its ores. Antimony, when an alloy with metals that are softer than itself, has the effect of hardening them. This property has been utilized in its most important alloy with lead, by which is made type-metal. It has the peculiarity of expanding rather than contracting when cooling; thus when cast into type the lines are made sharp, and, if the mold is perfect, the impression of the letter is clear and definite. The amount of antimony in type-metal varies; sometimes it is used in the proportion of 17 to 83 of lead or of 20 to 80, according to the kind of type to be produced. In stereotyped plates a small quantity of tin is introduced. Britannia-metal— which can be plated with silver so perfectly—is composed of antimony and tin; the proportion of the latter being 81 per cent to 16 of the former, with a very small percentage each of bismuth and copper.

Compounds of antimony are used extensively in medicinal preparations, that of tartar-emetic being one, and also in the manufacture of pigments for colors in painting, and in hardening or vulcanizing rubber. It is used as an alloy in quite different proportions with lead and tin. But we can not go into detail.

California, Nevada, and Utah Antimony.—The ores of antimony occur in abundance on the Pacific slope. The largest deposit yet discovered is in Kern County, California, and it was the first one of importance. This lode is immense, being thirty or forty feet wide; but the ore is uneven as to its value: some of it is rich and some is of a much lower grade. The vein consists of quartz and gray antimony, and is at an elevation from 5,000 to 5,800 feet, as though it had been pushed up from the great depths below. The quartz in connection with this vein is said to contain from sixteen to eighteen dollars' worth of gold to the ton in addition to the ore of antimony. In San Benito County, the ore found is of a specially high grade, averaging, it is reported, nearly 40 per cent. The ore (stibite of the mineralogist) of antimony is found in some half-dozen places in the State; in one of these it is associated with cinnabar, the ore of mercury or quicksilver, but upon the whole the number of these deposits of antimony appears to be very large. But the reader will remember that the industry of extracting antimony from native ores is yet in its infancy in this country, as, "up to the year 1883, but little metallic antimony had been produced in the United States." (*Min. Res. of U. S., 1883-1884, p. 650.*)

Nevada has numerous localities where ores of antimony occur. The one most prominent of these thus far discovered is in Humboldt County, a few miles south of the Central Pacific Railway.

Utah is credited with remarkable deposits of the ore, one of which covers 450 acres, and is estimated to yield 1,000 tons to the acre—"the purest antimony made, and promises to supply the world," while the ore gives an unusual percentage of metallic antimony. (*Resources of Utah, pp. 39, 53.*)

The ore lies in nearly horizontal, imbedded masses, within a sandstone formation, and at an elevation of about 6,500 feet. On the surfaces or slopes of these sandstone hills are often found large quantities of the ore, from which the attritions of the elements have worn away the once surrounding strata; sometimes these detached masses weigh from a few pounds up to tons.

Antimony, as we have seen, when alloyed with lead hardens the latter. In one instance—and others may yet be discovered—this has been done by the ores mingling in the mine. At Melrose, Alameda County, California, is an extensive deposit of these combined ores, and from which is extracted *hard lead*, equal to that made artificially, and which with equal success is used in antimonial alloys.

Bismuth.—The uses of bismuth are quite limited in extent and in importance. Some of these consist in making fusible alloys, as soft solder, and plugs for safety-valves to steam-boilers. It is used to a small extent in stereotype metal, and as an amalgam for silvering glass globes; in combination with a small portion of niter or saltpeter it makes the pearl-white enamel in porcelain, and with the same substance in a higher degree it is used to fix colors in dyeing. United with carbon or niter it is also employed to a small extent as medicine.

On the Atlantic slope its presence in Nature is little more than a trace. Thus, it is found in Monroe County, New York, and near Haddam, Connecticut; in Virginia, and in both North and South Carolina—but in no instance of commercial value, being merely of mineralogical interest.

Bismuth is found more abundantly in the West. In the vicinity of Beaver City, Utah, several veins occur amid limestone, the ore being quite pure, though it assays only from *one to seven* per cent of metal. In Colorado it is found in some half-dozen counties, as Hinsdale, Boulder, San Juan, and others. In the latter it appears united in a small degree with sulphur, and allied partially with a mixture of the ores of silver, copper, lead, and zinc. This combination of ores often furnishes beautiful crystal. Small but rare specimens of bismuth are sometimes found; these are usually very rich in the metal. It is known to occur in two places in Arizona, the one near Phœnix, the other near Tucson; the latter deposit is reported as containing a large quantity. Bismuth-ore of remarkable purity has been discovered, it is said, on a slope of Mount Vostovia, Alaska.

There is at present in the United States no mining of bismuth for commercial purposes, though it is thought by some experts that it may yet be utilized in extracting silver, and that will lead to "prospecting" for its ore. The small amount thus far produced from native ores has been obtained experimentally.

Arsenic.—Metallic arsenic is produced in only small quantities. It is a metal of a steel-gray color and brilliant luster, though usually dull from tarnish. It sometimes occurs in a native state, but it combines in the ores with several metals, as silver, iron, gold, nickel, cobalt, antimony, and also with sulphur. When highly heated, it rises in fumes like mercury or quicksilver, and thus, in smelting the ores with which it is combined, the intense heat causes much of the arsenic to pass off in fumes. Its cheapness and limited uses do not offer an inducement to collect it.

It is used as a hardening element in lead alloys; united with a small amount of sulphur, it is employed in making pigments, and also in giving colors in fire-works. The white arsenic of commerce is the common form, and from which are derived the arsenic compounds used in the arts. " In the gold and silver ores of the Rocky Mountains and the Pacific coast, arsenic is very common, but is regarded only as a hindrance in extracting these metals." (*Min. Res. of U. S., 1883–1884, p. 656.*)

Alum.—The alum in domestic use is chemically

obtained from certain clays of which it is a constituent, though sometimes it is found in small quantities in a native state, in which it is more or less impure, and requires refining before it can be used. Hitherto, American manufacturers of alum have depended upon imported material almost entirely, such as aluminous earths, alum clay, and bauxite (a clay impregnated also with iron), which are imported from England, Ireland, and France, and from Greenland—cryolite, the latter a peculiar substance, being a fluoride of sodium and aluminum.

Native Sources of Alum. — Clays that contain alum as a constituent occur in large quantities in New Jersey, Georgia, Tennessee, Alabama, Ohio, and Indiana, and in a smaller degree in some other States east of the Rocky Mountains. These clays when analyzed give from 15 to 40 per cent of alumina or sulphate of alum, while the foreign average between 45 and 50 per cent.

In Colorado small quantities of alum are found in several places, and also in Grayson County, Texas, while the Territories of Arizona, Utah, and New Mexico have alum deposits; the latter is pre-eminently rich in the sulphate of alumina. In California are a number of springs at the Geysers, the waters of which are impregnated with alum. At Sulphur Bank, Lake County, native alum is found in thick incrustations; in many of the placer mining-pits in the interior of the State it often crystallizes on the bed of the rocks laid bare in the process of hydraulic mining. But as yet little or no attempt has been made to utilize these alum deposits.

Aluminum.—Aluminum is comparatively one of the newly discovered metals, consequently its useful properties have become known only within recent years. This metal is indirectly the base of alum, the latter being the sulphate of alumina, whose metallic base is aluminum ; the latter is derived from the oxide of alumina, which is a constituent of certain clays. " Aluminum is a shining, white metal, having a shade between silver and platinum, and is very light, being only one fourth as heavy as silver of the same bulk." It does not oxidize or rust when exposed to moist or dry air, and, being quite sonorous, is often used for making bells. " Aluminum is of great value in mechanical dentistry, being light and strong and not affected by sulphur in the food." (*Knight's Mechanical Dictionary.*) This metal is used largely in the manufacture of cheap jewelry, in the form of an alloy with copper—that is, ten parts of aluminum to ninety of copper. This alloy is a hard, pale, gold-colored material ; it is harder than bronze, and takes a fine polish which scarcely tarnishes. A small quantity of aluminum in the form of an alloy of iron, when put in a crucible of melted iron, causes the latter to flow easily. (*Min. Res. of U. S.*, *1885, p. 392.*) Aluminum bronze is available for many uses, as druggist's weights, thimbles, table-furniture, hardware, door knobs and handles, chandeliers, busts, statuettes, vases, and numberless other purposes, among which are the works of watches. It also makes alloys with silver, iron, nickel, and zinc. It may be remarked that the cap or apex of the Washington Monument, at the Nation's capital, is pure aluminum. It was made by Colonel William Frishmuth, of Philadelphia, and is in the form of a four-sided pyramid, ten inches high, each side at the base being six inches. " Until recently the aluminum sold in the United States was entirely of foreign origin, but it is now produced in this country by a process patented by Colonel Frishmnth." (*Min. Res. of U. S., 1883–1884, p. 658.*)

This process for obtaining aluminum is said to be much less expensive and more effective than that which is obtained abroad.

Deposit of Alum-Rock.—The indications are that the United States will soon be able from their own resources to supply aluminum sufficient for the wants of the people. A deposit of " the sulphate of alumina," was discovered in 1884 in Grant County, in Southwestern New Mexico. This immense deposit is on the upper course of the Gila River, about forty miles from Silver City. " It covers an area of 1,600 acres, and may be termed a mountain of that material, or alum-rock. It presents as to color a pink, gray, and yellow-tinged substance, containing an extensive crystallized formation of what proved after chemical examination to be almost *pure* sulphate of alumina, with traces of iron. Some of the cliffs rise to a height of 800 feet above the bed of the river, and present faces of this substance, hundreds of feet square in extent. The whole is the result of volcanic upheaval, as clearly indicated by the lava-rock that surrounds it." Four samples taken from different parts of the deposit and analyzed by a competent chemist—Dr. Charles J. Bandman—give as an average 14.04 per cent of oxide of aluminum, and of sulphate of alumina 40.14. The deposit is practically inexhaustible and is waiting to be utilized. The location is within forty miles of the railway at Silver City, but it is in the midst of deep forests and rugged mountains.

Mica.—Mica is a mineral that is scaly in its structure, so that it can be smoothly split into elastic sheets of great thinness; it is quite translucent, and the very pure is nearly transparent. Once it was known as " Muscovite glass," being used as such by the old Russians. It is common in connection with granite and some other kinds of rocks, as it is an " essential constituent of granite, gneiss, and mica slate." In appearance when in a pure state it is almost colorless, but often bordering on gray or light green, while the impure has shades of a darker hue. Its value is greatly enhanced when in sheets of sufficient size to be available, and being virtually incombustible it is used extensively in the doors of stoves, also for lamp-shades and lanterns. The very finest is often used for dial-plates in compasses. An immense wastage is incurred in the trimming of the sheets for market; this waste is somewhat utilized, when it is

ground to minute particles, and used in the manufacture of certain styles of wall-paper, in giving it a
glistening effect. These are only a few of the uses
to which it is applied.

Mica ; where found.—A mica belt extends along
the Atlantic slope, from Maine to the extreme southern end of the Alleghanies. " It is found in bunches or pockets, in connection with large masses of
quartz or huge crystallizations of feldspar." With
mica are often associated rare and valuable minerals,
such as beryl, tourmaline, garnets, etc.

In Maine rich deposits occur in a number of
places, such as at Mount Mica, where excellent
sheets have been obtained, and where have been
found an unusual number of the " famous gem tourmaline." In New Hampshire, southwest of the
White Mountains, commences a mica belt that extends in a southerly direction. " Throughout that
region the granitic veins are most conspicuous ob-
jects, and often they may be seen from miles away
cropping out along the hill-sides." Here at intervals
are deposits of mica. In one locality—Alstead—
sheets have been mined that were nearly four feet
across. This mine has been worked forty years,
while another in the same belt has been for more
than three fourths of a century ; while there are also
large numbers of similar deposits more or less extensive within the same limits. Immediately south in
the States of Massachusetts and Connecticut the belt
extends within similar granitic veins. But in the lat-

ter State are instances where the mica deposits, as at Glastonbury and other places, are worked more to obtain feldspar than for mica.

In the States of New Jersey and Pennsylvania discoveries of mica have been made, and also a few isolated attempts at mining. But a remarkable deposit at South Mountain in the latter State is said to have produced pieces of mica weighing from nineteen to twenty-seven pounds, some of which " could be split 160 times to the inch." (*Min. Res. of U. S., 1882, p. 583.*)

Maryland has a number of mica deposits in Howard and Montgomery Counties. These mines are more or less scattered within that region, one of which is twelve miles north of the city of Washington. The belt crosses over into Virginia, and mines have been worked to some extent in several localities along the belt line. Near Amelia Court-House it is associated with the mineral beryl. Mica is found in South Carolina, Georgia, and Eastern Alabama, but as yet mines have scarcely been opened.

North Carolina Mica.—This State contains the richest portion of the Alleghany mica belt, and here the industry of mining it has attained the greatest success. The deposits are numerous in the highlands or mountains in the western part of the State; the counties in which the mines are the richest, and the production greatest, are Yancey, Mitchell, and Macon. " Mica is found in ledges [veins] of very coarse granite. Many of the plates are of remarkable size, reaching three and even four feet in diameter. . . . This region furnishes the bulk of this mineral to the world's markets." (*Handbook of North Carolina, pp. 179, 180.*)

Rocky Mountain Mica.—When we leave the Alleghanies for the West, we meet with only one deposit of mica, and that is in the Northern Peninsula of the State of Michigan, till we reach the Black Hills of Dakota. Here amid granite of very coarse crystallization mica occurs, and here it is mined in unusually large sheets; in this region in almost every instance it is associated with tin (p. 230). While the indications are that this entire region is rich in mica as to quantity, it is no less remarkable for its clearness and perfection in texture. In Wyoming Territory it occurs plentifully in the Wind River country, and in several places along the mountain-ranges in Laramie County. The grade is very fine, being almost transparent, and the sheets quite large. In Idaho, in the Cœur d'Alene district, mica also abounds.

In Colorado, thirty-five miles south of Denver, is a deposit of mica of fair quality and quite extensive in size, while it is reported as having been discovered in other portions of the State. New Mexico has also mica-mines in the vicinity of Las Vegas, the capital of San Miguel County, and also in several other localities, but as yet undeveloped. Arizona has deposits likewise, and Nevada has the credit of possessing several deposits of salable mica, all ready for future mining and use.

California has deposits in nearly every county in the State; it is said the most important of these are in the Salmon Mountains, in the northwestern part of the State. Oregon and Alaska have also their share of mica, but held in reserve for future use.

The reason that these numerous deposits of mica, in such diverse localities in the Union, are not more fully developed, is, that comparatively little of it is needed to supply the wants of the people at the present time; but in the future there may be discovered other uses for it of which we now little dream.

Asbestus. — This mineral is a variety of hornblende, and is unaffected by fire. It often occurs in long, delicate fibers, or fibrous masses, usually of a white or gray color, but sometimes greenish or reddish. The finer varieties have been wrought into gloves and incombustible cloth, while the longer is sometimes used in making ropes. It is found in the United States in very many localities, but always in small, isolated deposits or pockets; in consequence, only a limited number of these are mined, so that in 1884 (*Min. Res. of U. S.*), in the whole Union, only about 500 short tons were taken out, and those from numerous places. It occurs occasionally in Massachusetts, as in Brighton, Pelham, and Windsor; in Richmond County, New York; near Brunswick, New Jersey; and farther south in small deposits, here and there, along·the Atlantic slope, from Pennsylvania to Georgia. On the Pacific slope it is found under similar conditions, and is reported to exist

18

in Dakota, Wyoming, Colorado, Utah, and Nevada. The American, thus far discovered, is usually of a short fiber, being also somewhat brittle and harsh. This class is used in the manufacture of fire-proof safes, and also in that of fire-proof paints, cement, for steam-packing, fire-proof enamel for walls, and drop-curtains in theatres. It has also been introduced for the purpose of insulating electric wires.

XXV.

ANY intelligent person, enumerating the natural resources of the United States, will scarcely fail to notice their peculiar features—their wonderful variety; their immensity and presence in so many localities; their universal and practical use in promoting the people's comfort and material progress, as if they were intended to make the various sections of the country dependent upon one another, and induce the people themselves to cherish the interests of the entire Union as mutual. When we compare our native precious stones as a natural resource, how great is the contrast with other countries! How meager they are, both in number and quality, and in ostensible value! In the Old World diamonds and other jewels, of great size and pure quality, are used only for show and ostentation, and valued as such by royal families or heads of empire; the finest diamonds and gems in the world are to-day "crown-jewels."

When taken in the aggregate, we learn the comparative value of the thirty-one varieties of American native precious stones, sold in one year (1884) to be used as jewelry, and, for the same year, the value of the precious stones imported—the former was $28,650 and the latter (not set) $8,712,315. Among the native American stones, the diamonds were valued at only $800.

Diamonds.—The most precious of stones, the diamond, is found in the United States in very limited quantities, of small size and seldom of first water. They have been picked up occasionally, in several and widely separated localities, but in no place have they occurred in sufficient numbers to warrant for them an extended search. Many small specimens have been found in gravel-beds or in alluvial soil. One of these is represented as very beautiful and of first water or quality; yet it was worth only a hundred dollars. Another was of a yellowish color, another of greenish, and still another was large but of a dark color; the latter was destroyed by ignorant laborers who were trying to break it: the fragments proved it to be a diamond.

Geologists have noticed that itacolumite—"flexible sandstone"—which has been regarded as the *matrix* of the diamond, occurs in certain localities in South and in North Carolina, and in Georgia; and, in the vicinity of this flexible sandstone, numerous small specimens of the diamond have been obtained. " The series or group" (of flexible sandstone) "is an interesting one, from its supposed relation to the diamonds that have been found in this State and in

South Carolina as well as in North Carolina." (*Commonwealth of Georgia, pp. 80, 140.*) The largest native diamond thus far known in the United States was found by a laborer when digging in the earth at Manchester, Virginia, near the James River. This gem, being a little off color, is valued at only $400. There have been numerous instances in which diamonds have been found in the Union, as in California and Nevada. The latter have all been obtained in gravels or gold-washings; but these afford no clew as to where they originated.

Miscellaneous Stones.—There are known to exist in the United States more than a hundred varieties of what are termed precious stones. About ninety of these are susceptible of receiving a polish, which renders them suitable as ornaments; they are also of numerous shades of color. We will notice, but very briefly, the better known and more valuable of these "stones," their characteristics, and where found.

Emerald and Beryl.—These two are almost identical, the difference being in the origin of their color: the first by the oxide of chrome, the second by the oxide of iron. Their colors vary from a lively, beautiful green to a dark green, while some are rose-color and greenish yellow, and even deep blue. They are found in a number of places in Maine, in Connecticut, in Massachusetts, and in Pennsylvania, but much more frequently in North Carolina.

Hiddenite.—In Alexander County, North Carolina, and thus far nowhere else, is found an *emerald-green* gem known to jewelers as Hiddenite, from the name of the discoverer—W. E. Hidden. The crystals are found in nests or pockets, but not in veins. In these small deposits are also often found emeralds, and beryls in connection with this lately (1881) discovered jewel. "Hiddenite is the only strictly American gem." This stoné has been mined for by sinking shafts, and, because of its unusual beauty and rareness, is in great demand, far beyond the supply. "It is found in the form of slender crystals, having emerald color, but totally different in all other respects from emerald proper." It possesses "hardness, beauty, and brilliancy," while its fine color is derived from the oxide of chromium; its basic element is *lithia,* but that of the emerald is *glucina,* while both are silicates of alumina. (*Mr. Hidden, in Southern Geologist, and Handbook of North Carolina, p. 201.*)

Topaz is found in Maine, in Arizona, New Mexico, and in Southern Colorado at Pike's Peak. At the latter place fine specimens have recently been found; they are of a beautiful light blue color and quite clear. These are deemed equal to those of the same size from Siberia. Utah has topaz crystals, often of a wine-color and yellow, but generally limpid white; they are usually small.

Sapphires and Rubies.—These are closely allied with corundum. They occur in New Jersey, and in Macon County, North Carolina, where about fifty gems were found. "The colors were rich blue, violet-blue, ruby-red, light red, pink, and yellow, and others were colorless." The richest locality for sapphires in the United States is near Helena, Montana, where they are associated with garnets. Near Santa Fé, New Mexico, in Southern Colorado, and in Arizona, they are found in the sands; in color, sapphire, blue, or ruby-red.

Garnets.—Garnets of moderate gem value occur in New Hampshire. Fine specimens have been also obtained at Round Mountain, in Maine, and near Avondale, in Pennsylvania, and at Stony Point, North Carolina. The finest ones for gems are from New Mexico, Arizona, Colorado, and Nevada. These are called "rubies," the blood-red color being the favorite, with the hyacinth-yellow; they have the peculiarity of appearing specially beautiful in artificial light. Garnets are also found near Fort Wrangel, Alaska.

Tourmaline.—The richest locality for this crystal thus far known is in Androscoggin County, Maine. These specimens are usually colorless, but often light pink, light blue, bluish pink, light golden, and in some other localities the colors are green and red. When cut as gems, some of these colors assume a darker hue. Tourmaline is sparingly found in New York, and in Delaware County, Pennsylvania.

Obsidian.—This is a kind of glass produced by volcanic influence. It occurs in the United States within the range of the Rocky Mountains and in their vicinity. Numbers of a smoky, transparent variety with a greenish tinge, have been found near Santa Fé, New Mexico. It is in the form of rounded pebbles, often an inch in diameter. Another variety has been discovered on Gunnison River, Colorado; and still another, of light gray and clear, concentric structure, has been found near Georgetown in the same State. Just across the line in California, in Owen Valley, occurs an obsidian of a red color, banded with alternate layers of black and brown. Fine varieties and very abundant are also found near Silver Peak, Nevada. Yellowstone Park contains obsidian, often in fine black specimens, and others mottled black and brown, in small layers. American obsidian is but little used as jewelry.

Amethyst.—This stone is usually of a light purple color bordering on a pink, but it has comparatively little value as a gem. Amethysts are found in numerous places in New England and in Delaware and Chester Counties, Pennsylvania. Near Clayton, Georgia, have been obtained some rare specimens. " At times these have within them large liquid cavities containing movable bubbles of gas." This stone occurs more in the West than in the East. " The Lake Superior variety is spotted with the coating of red, moss-like markings, giving them if cut a moss-amethyst effect." In the Yellowstone National Park " amethysts line the hollow trunks of *agatized* trees, varying in color from a light pink to a dark purple." In the main valley of the Yellowstone are thousands of silicified trees. " In some cases the structure of the tree is well preserved, and in other cases agatized or opalized, and lined with crystals of beautiful amethysts." Wisconsin, Texas, Nevada, and Colorado abound in amethysts.

Agate and Chalcedony.—There is an intimate connection between agate and chalcedony, both being uncrystallized quartz, but "the colors of the agate are delicately arranged in stripes or bands, or are blended in clouds," while the "chalcedony is translucent and usually of a whitish color, and of a luster nearly like wax." When the stripes of the agate are parallel, it is known as *onyx;* when the chalcedony is flesh-red, it is called *carnelian.* Agates are found in many places along the Connecticut River; in Delaware and Chester Counties, Pennsylvania; and in a number of localities in North Carolina. An unusual abundance of agate in all its varieties occurs on the shores around Lake Superior and on the upper Mississippi. Many very beautiful specimens are obtained in different parts of Colorado and throughout the Rocky Mountains, in Arizona, New Mexico, and in California. The *moss-agate* variety occurs in North Carolina and in Pennsylvania, and in large numbers in Humboldt County, Nevada.

Serpentine (Jeweler's). — This stone is often greenish in color, but mottled with shades like a serpent's skin. "The finer varieties are translucent and of different shades of rich oil-green color, usually dark, but sometimes pale" (Prof. Dana). These grades are designated "precious," and jewelers cut them into various forms according to taste, and as such they are worn as ornaments. Beautiful specimens of a dark-green color, but in limited numbers, are found in the vicinity of Newburyport, Massachusetts; while others, whose prevailing color is yellow, occur at Montville, in the same State. These stones, but small in numbers, also occur in the States of Maine, New York, New Jersey, Pennsylvania, and Maryland, and in quite large numbers in North Carolina, but of a low grade of texture.

Jade of a dark-green color has been obtained from the natives in Alaska. It is supposed to exist in places somewhere to the east of Point Barrow.

Opal.—Fine opal, without any opalescence or play of colors, is found in Georgia and in North Carolina; and common opal, of greenish or yellowish white, with vitreous luster, occurs in Lebanon County, Pennsylvania; also in Colorado and in many places on the Pacific slope, and in Idaho, have been found "specimens showing play of colors." (*Min. Res. of U. S., 1882–1884, Precious Stones.*)

Jasper.—This stone is of various colors, and is found in many localities in the Union. It is an opaque, impure variety of quartz of red, yellow, green, and other colors, and is susceptible of a polish. It is marked sometimes by regular stripes. This stone has been found in North Carolina in several localities, having the texture known to jewelers as " cat's-eyes," and also in the same form in Rhode Island. Jasper occurs in Kansas, California, and an unusually large bed has recently been reported in Arizona. (*Min. Res. of U. S., 1884, p. 761.*) This stone was once used as a gem, but has passed out of fashion, and "is very little used in the arts for so common a stone."

Jet.—This is not a stone, but a variety of lignite or mineral coal. It is of compact texture and velvety black color, and withal susceptible of a good polish. It is often used as "mourning ornaments," but as such is nearly superseded by black onyx. It is found in quite large numbers in the lignite "coal-bearing rocks of Colorado."

Amazon-Stone.—This is a variety of feldspar of a verdigris-green color, found originally near the Amazon River—hence the name; it is opaque, but takes a fine polish. This stone has recently been found at Pike's Peak, Colorado, and it has a wide reputation for its rich green-colored crystals, though it is not used extensively as a gem.

Turquoise.—A Persian stone that found its way into Europe through Turkey—hence the name. In color it is of a peculiar bluish-green, and is susceptible of a very fine polish. Within recent years it has been discovered in several places in New Mexico, at Mineral Park, Arizona, and also in Nevada. The color of the gems found in Arizona is nearly all of an apple- or pea-green shade, though occasionally blue. The estimated gem-value of the turquoise is not very high.

XXVI.

FROM time immemorial man has used clays in forming useful vessels, and in the course of generations he discovered and utilized the various grades that make the different kinds of ware, from the commonest to the most delicate porcelain. The deposits of clay from which the ordinary brick is made are so numerous and common throughout the Union that they may be classed almost as universal. The clays of these beds often differ in grade: while some are suited only to the manufacture of bricks having a coarse texture, others are susceptible of producing those of a finer and smoother quality, and also common earthenware and terra-cotta. The deposits of these clays are so widely distributed and so accessible, that in respect to them further detail is unnecessary.

Fire-Clay.—The quality in clay that enables it to resist heat is termed refractory ; this is the valuable element in fire-clays. These have different grades in their texture : some are coarse and harsh, and some fine and smooth. The former are intimately asso-elated with the bituminous coal-measures, and are found underlying the seams; the latter occur outside these measures, and have apparently a different origin. We shall not further notice the former, as the reader will understand that beds of this clay almost everywhere underlie the seams of coal. The clay derived from these beds is specially adapted for manufacturing fire-bricks, which are used in lining ranges, stoves, walls of furnaces, etc., where great heat is required. In addition, this clay is also utilized in making drainage-pipe, stoneware, and numerous other useful domestic articles. We shall now treat of the clays outside the coal-measures.

Massachusetts has deposits of fire-clay at Martha's Vineyard, and Vermont an extensive bed of the same near Brandon, and also at Bennington; Connecticut has kaolinite in Litchfield County, and New York, in Dutchess County; but New Jersey has thus far the credit of possessing the most extensive deposits of excellent fire-clay in the Union. This belt commences in the eastern middle of the State and extends in a southwesterly direction from the vicinity of Perth Amboy to Trenton, and thence below on the Delaware. The entire area of the belt is 320 square miles. This formation is composed of "a series of strata of fire-clay, potter's clay, brick-clay, sand and lignitic clay." (*Min. Res. of U. S.,* *1882, p. 465.*) This State produces about three fourths of all the articles made from that clay in the United States. Passing southward, we find Pennsylvania having numerous and extensive beds of fire-clay, among which is an eight-foot vein at Saltsburg, recently discovered. Maryland and Virginia, and a portion of North Carolina, have fire-clays in many places, but all associated with coal-measures. A deposit of fire-clay is reported (1885) at Evansville, Indiana, and "a fine pottery-clay in Lincoln Parish, Louisiana."

Kaolin.—This clay is defined as the decomposition of feldspar, which is composed of "silica, alumina, and potash." The name is derived from a Chinese word—*kaoling*—designating the clay from which they make their famous china-ware. In North Carolina kaolin is found abundantly from one end of the State to the other; valuable for making china and other wares, for paper-manufacturing, and for fire-brick. . . . "This kaolin, or white earth, had been exposed, like snow-banks, in huge dumps and open cuts by an ancient people, the mound-builders, a thousand or two years ago." (*Handbook of North Carolina, p. 182.*) It is interesting to note that more than two hundred years ago a quantity of this clay was sent to England, where great exertions were in progress to discover somewhere in the king's dominions a clay similar to that from which the Chinese made their delicate ware, just then introduced into Europe. Also in South Carolina, in the vicinity of Aiken Court-House, "large beds of kaolin-clay, free from grit or other impurities and of great whiteness, are found intercalated, or between the layers of sands

. . . this material being used as porcelain-clay and for glazing by paper-manufacturers," and in a number of other places in the State. Some of these clays are beautifully mottled with various colors, and harden when exposed to the atmosphere. (*Resources of South Carolina, p. 120.*) Extensive beds of white porcelain and pottery clays occur in Georgia in a number of places, and common-ware clays in the Tennessee Valley and in Kentucky. In the northeastern part of the State of Mississippi yellow and cream-colored clays are common in the sand-formation; in Texas, they are in connection with coal-measures, while " a large deposit of kaolin produced by the decomposition of granite " occurs in Pulaski County, Arkansas, and beds of pottery-clay in other places ; the latter is also found near the center of the State of Louisiana.

Mississippi Valley Clays.—Minnesota has an extensive deposit of kaolin derived from decomposed granite, and Wisconsin has an area of about 750 square miles where beds of kaolin are found. Indiana is the fortunate possessor of fine porcelain-clays, which are used extensively in manufacturing white ware and also for "encaustic tiles of rare beauty and excellence, equally vitrified and as good as those produced in the best factories of England and France." Encaustic or enameled tiles of equal grade are made from porcelain or pottery clays obtained in the vicinity of Zanesville, Ohio, "but parts of the mixture are from abroad." These glazed tiles are used for hearths, halls, vestibules in large buildings, as churches, etc., and for miscellaneous ornamentation. Some of these clays have flint or silica as an ingredient, and are utilized in making various articles, such as drain-pipe and common earthenware.

Missouri has several valuable deposits of kaolin and an excellent fire-clay, especially in the counties of St. Louis and Montgomery. The latter clay is valuable, as from it are made pots used in glass-making. These pots are carefully constructed, or "built," as the workmen say. In them are put the ingredients of which the glass is to be composed, and the pots are subjected to intense heat in order to melt the mixture. There are a number of beds of kaolin in that State. "This material has been thoroughly tested, and from it has been produced elegant porcelain-ware." In Jefferson County is found . a white clay, known as ball clay; in one place it is practically inexhaustible; from this are manufactured queen's-ware and other articles of use. (*Handbook of Missouri, article Clays.*)

Rocky Mountain Clays.—Fire-clay is found in Wyoming Territory, along the Laramie and Wind River Valleys, and in various other localities. Near Helena, Montana, are deposits of fire-clay, but connected with the lignite coal-measures. "It is found in many places in great purity, and in extensive beds, and suitable for a large number of purposes, the supply being abundant." Utah has, in the vicinity of Salt Lake City, several deposits of fire-clay; and in

California and Oregon are numerous beds of kaolin, but the quality of the article is not of the finest. These have been somewhat utilized, but the much greater portion remains undeveloped. " In Washington Territory remarkable specimens of kaolin have been found; and clay suitable for making fire-brick and the ordinary brick for building is plentiful and well distributed." (*Report of the Governor, 1886, p. 32.*)

Porcelain.—In conclusion, it is proper to state that certain portions of the kaolin found in the Union is of a superior quality. Prof. F. ʌ. Wilber, in the "Mineral Resources of the United States," speaks of ware produced from American kaolin, " Parian porcelain," made only in this country, and which attracted attention in England because of its excellent qualities and beauty, and also of a delicate form known as "egg-shell porcelain," and of a "porcelain paste" as pure as old Sèvres, the porcelain made from which has "a characteristic creamy tint, and is very translucent."

GLASS MATERIALS.

A number of ingredients are mixed in the material from which glass is manufactured; the combinations of these ingredients are the results of experiments made during hundreds of years. An essential one in every kind of glass is silica—an element of flint—in the form of sand. To this are added soda, lime, potash, and often a number of metallic oxides, according to the kind of glass desired. The various grades and kinds of glass are obtained by using certain proportions of the ingredients in the admixture —or "batch," as the workmen term it—in the pot where the material is melted. An abundance of all the numerous materials used in manufacturing glass is found in different localities all over the Union; they are excellent in quality, and practically unlimited in quantity. "No better glass-sands are found than those of the United States. . . . Some of the most beautiful colored glass produced in the world, rivaling, in depth and richness of coloring and beauty of design, that of the famous works of Europe, is made at the flint-glass works of the United States." (*Min. Res. of U. S., 1885, p. 544.*) Some of the works in Pittsburg and vicinity have introduced natural gas as a heating power, and with the effect of greatly improving the glass.

LIME AND CEMENT.

In the construction of brick or stone walls for buildings, or for other purposes, it is essential that lime mixed with sand should be used as mortar, but often under certain conditions cement is also necessary. The former is obtained by burning common limestone, a natural resource in such quantities and so widely distributed throughout the Union that the very numerous localities where it is found need no further notice in this volume. Cement in the main is also derived by burning a grade of limestone that exists very extensively in the United States, and in widely separated localities, extending from the Atlantic to the Pacific. Common limestone is a carbonate of calcium—the metallic base of lime; while hydraulic limestone has in its composition also *silica* —the chief element in flint—and more than a half-dozen other ingredients. From this combination is manufactured hydraulic cement or mortar. The latter from these ingredients derives the property of hardening speedily, of being insoluble, and is used in cementing under water. Where great strength and tenacity are required, as in the foundations of heavy buildings, in cellar-floors to prevent water penetrating, in sewers, and aqueducts, etc., and often in the manufacture of artificial stone, and in solid foundations for pavements, cement is in universal requisition.

Hydraulic limestone is found in very numerous places in the Union. Of these, the two most extensive deposits thus far discovered, one is in the valley of Rondout Creek, Ulster County, New York; the other in the vicinity of the Falls of the Ohio; the latter extends on both sides of the river into the States of Kentucky and Indiana: here, as in New York State, are large establishments for preparing cement. The materials for making cement are abundant in the State of Tennessee, and they are found in quantities in Colorado, and on the Pacific slope; in California, in the State of Oregon, and in Washington Territory.

XXVII

OUR building-materials, as furnished direct from Nature, are great in number and in variety. The stones to choose from are of different kinds in texture, but having those qualities that can withstand the influence of climate, in not being easily disintegrated nor disfigured by change of color; while clays suitable for making bricks of various grades are available in every section of the country. The main divisions of our building-stones are reckoned in three classes : 1. The entire series under the common name of granite, with their varied forms of texture and shades of color and susceptibility of polish. 2. The series whose principal ingredient is lime, as the various kinds of limestone, including marble of different and coarser grades. 3. The sandstones of several colors, and of coarse or fine texture.

Characteristics and Uses of Stones.—We can treat of these only in general terms when noticing the classes of stones and the localities where they are found. Granite is defined as " a crystalline, unstratified, and true igneous rock." When polished it usually appears of a whitish, grayish, or mottled with a flesh-red color, and sometimes quite dark. This stone is seldom used for private edifices, but generally for large and imposing public buildings. The United States Government uses it almost entirely for its massive structures. The coarse kinds are often used in the humble capacity of pavements in the streets of our cities wherein there is a large amount of heavy traffic. Owing also to its enduring qualities, it is rapidly superseding marble, from the simplest to the most imposing monuments, in cemeteries. It is noticeable that " brick in a general way, pressed brick, and brick and terra-cotta work combined, especially on the Atlantic seaboard, have somewhat superseded stone."

Marble.—Marble as a building-material appears to be gradually diminishing in use in portions of the country—notably in New York city and vicinity—owing perhaps to its tendency to discolor, so that in a few years its brightness becomes dim. Vermont marbles, " dressed to a harder, smoother, and durable surface," are used in trimmings, especially where the walls are of brick, and for interior ornamentation. In accordance with the fancy of the architect, if supplemented by the purse of the owner, building-stones of various qualities, and from widely separated localities, are often transported long distances to be used in the same structure. For illustration, houses in San Francisco have been built of Vermont and Tennessean marble, and in New York city of oölitic limestone from the interior of Kentucky; while in the way of trimmings our cities have no hesitancy in laying under contribution any State in the Union, if it has the special quality of stone wanted; for instance, New York uses marble from Tennessee and Georgia side by side with granite from Maine, or even from that outsider, Nova Scotia.

Sandstones.—The brown sandstone appears in some respects to be the favorite with the majority of owners of private houses, who adopt that class of stone for fronts, as, when of good quality, it resists successfully the effects of heat and cold, dryness and moisture; the best grade being so compact as not to admit the moisture within its texture, as the former will freeze in the cold and cause the stone to scale off. Competent architects understand in what position to place them in the wall; it being proved that the best is that in which the stones originally lay in the quarry. Sandstones of this class are quite expensive to construct walls with them alone; in consequence, they are used only for the fronts, the latter being often a few inches in thickness, but anchored and thoroughly backed up by common brick—the brown-stone being used as a sort of veneer. This material is unusually popular for church edifices in the cities and villages that are within reach of its famous quarries in New Jersey and Connecticut.

There are several grades of sandstone - rock which are available for building purposes. These partake of one general property, that of sand being well and firmly compacted; but this sand is often quite differently constituted. Geology tells us that in this class of stone the basic sand may be silicious, or flinty, or granitic (derived from decomposed granite), argillaceous, or clayey. This accounts for the various grades of sandstone : some are gritty, some savor of the granite, and others partake more of clayey properties. Then, again, we find different shades of color, as brown, some darker than others; as drab, some lighter than others ; and some partaking of a bluish, slaty tint. We have also in immense quantities, scattered in places throughout the Union, another class of building-stone, the basis of which is lime, such as the common limestone and gypsum, and the coarser grades of marble.

Localities of Building-Stones.—Granite predominates in the New England States, and in them are found, when taken all in all, the finest specimens of that stone in the Union, there being several varieties in respect to hardness, texture, and color. Portions of the Maine granite when polished have an appearance that reminds one of the scales on a shad; the Quincy has quite a dark luster; that of Cape Ann still different; and in New Hampshire, the " Granite State," are found several shades of color in the polished stone, some almost white, though speckled, and some of a grayish tint; the Connecticut shows reddish specks amid flakes of gray ; while Rhode Island granite is excellent, and usually of a darkish gray in color. These granites in their varieties are all beautiful and possess a remarkable solidity. Because of these two qualities the United States Government has used New England granite in building nearly all its massive structures on the Atlantic seaboard, and also in many of the great public edifices in the city of Washington. In the latter, however, granite from Virginia and North Carolina has been sometimes used.

The Granite of Five States.—There exists a great abundance of building-materials in these States. Granite and gneiss are among the commonest rocks in the western highlands of North Carolina, and these deposits border on the extensive granitic region of East Tennessee. The former State has about eighty varieties of building-stones, among which is claimed a granite similar to the famous Scotch. (*Handbook of North Carolina, p. 197.*)

The rocks of the upper country of South Carolina are similar to those just mentioned. In Newberry and Fairfield Counties are extensive beds of excellent granite, fine-grained and easily splitting, inexhaustible quantities of the best building granite, and in many other localities of the State. One area of several square miles " furnishes the finest quality of blue and white granite. Granite occurs in almost unlimited quantities in the vicinity of Columbia, the State capital; the rock is of a light-gray color, fine-grained, compact, is durable and of uniform texture." (*Handbook of South Carolina, pp. 131, 608.*)

Two or three classes of granite are found in Middle Georgia ; they are distinguished more or less by their colors: some are flesh-colored in specks, while others are gray, and some dark in their luster. Rocks having the granitic properties, such as gneiss, are very common in the highlands of the State. Iowa also has granite in a limited extent in some of the northern counties, but it does not lie in regular beds so that it can be quarried, but is found in huge bowlders. From this granite is built the State Capitol. An extensive deposit of granite exists in the valley of the Colorado, in Burnet County, Texas. This granite is similar in its main characteristics to the famous red syenite of Egypt, of which the obelisks are made, and also to the Scottish red granite. The State Capitol at Austin is constructed of this material.

Rocky Mountain Granite. — Building - stones of almost every variety, including " volcanic rock," are found in this division. Granite occurs in great quantities in Montana and Wyoming, and in a less amount in Dakota. In the first two there are several varieties partaking of nearly all colors and intermediate shades: some are coarse and some of fine grain, the latter being susceptible of a good polish. In Colorado building-materials are in abundance, and of course have been brought more into use than in the Territories just mentioned. The extensive deposits of granite in the State are of different varieties ; some being of red crystalline rock, which is susceptible of a high polish, making a fine material for ornamental work, while there are other granites of a light-straw or cream color, that take a fine, durable polish. Colorado has also an abundance of a favorite building-stone, a fine *pink*-colored lava, which is used quite extensively in the city of Denver.

Pacific Slope Granite. — An untold amount of granite is found within this region amid the foot-hills of the Sierra Nevada, while the latter is one immense mass of granite from end to end. Large amounts of this substantial stone are utilized in San Francisco, in the construction of public buildings and in the United States dry-dock at the Navy-Yard. It is a singular fact that the first granite used here for building purposes came from China, dressed and ready to be put in the wall. In some instances the granite of this State is very compact. In the Penryn quarry, east of Sacramento, the stone splits so evenly that blocks 100 feet in length, and of almost any thickness required, have been taken out. This granite does not change color by exposure to the atmosphere, hence it is adopted for monuments or walls that are exposed to the elements. " The predominating shades of Penryn granite are blue, gray, and black ; the last named very much resembling the celebrated black granite found in Egypt. and they are exceedingly beautiful when highly polished." (*Min. Res. of U. S., 1882, p. 455.*) California has abundance of granite in numerous other locali-· ties, and the State of Oregon and Washington

Territory have each granite in their eastern portions.

Building-Stones **whose Base is Lime.**—These stones are of three classes—common limestone, gypsum, and a marble of a comparatively coarse grain. Perhaps the most extensive deposit of the latter, yet discovered in the Union, occurs in Westchester County, immediately north of the city of New York. From this source has come almost all the marble used in the city for buildings; the most prominent structures of this marble are the New York University and the Roman Catholic Cathedral, the latter the most imposing church edifice in the Union. The main towers are, however, constructed of a white, hard, and beautifully coarse marble quarried in the vicinity of Baltimore, Maryland. There are several other localities within the State where limestone of a fine texture is obtained, and which is adapted to building purposes. At Lockport, on the Erie Canal, are extensive beds of such stone. The latter when taken from the quarry is comparatively soft and easily worked, the blocks being dressed and so jointed that they can be placed in walls with ease. The fine texture of this limestone and its imperishable nature (for when exposed to the atmosphere it becomes exceedingly hard, and its color, a delicate gray, pleasing to the eye) have rendered it a favorite. The Lenox Library building, on Central Park and Fifth Avenue, in New York city, is constructed of this beautiful stone.

On the Atlantic slope, North Carolina comes next to New York in having, in the western portion of the State, extensive beds of crystalline limestone—a coarse marble well adapted for building purposes; while in the coast region "shell limestone" abounds as a building-stone. In East Tennessee, just on the State line, is found an abundance of oölitic limestone —characterized by round grains as small as the roe of a fish, but of excellent quality. It is very white, works easily, and stands exposure to the weather very well. Vast beds of the same kind of limestone are found in the vicinity of Bowling Green, Kentucky. Throughout Middle Tennessee, limestone is available, and extensively used for foundations. Some of these beds are almost marble, as in Henry County. The State Capitol at Nashville—so imposing in its general features—is constructed of this limestone.

Let it suffice to say that limestone is common throughout the valley of the Mississippi, but in the main, from its general nature, it is used for other purposes, such as making lime, rather than for buildings; to this, however, are found a few exceptions. Some portions of this middle region of the Union are furnished with another class of building-stone—gypsum —in which lime is one of the constituents; the difference being that limestone is a *carbonate* of lime, and gypsum a *sulphate*. Gypsum is found in immense and inexhaustible quantities in the State of Michigan, but in that connection it is more valuable for other purposes (p. 472).

Iowa has an abundance of limestone, for building and for obtaining lime, in many counties of the State. The finest and largest deposit of gypsum in the State is near Fort Dodge; it extends in solid rock formations for five miles along the Des Moines River. This stone is used for building purposes; it is quarried like limestone, and is capable of being cut into blocks of any shape or suitable dimensions. (*Min. Res. of Iowa, pp. 72-74.*)

Missouri has great varieties of excellent limestone throughout all sections of the State; these deposits supply fine building-stones and are used as such extensively in all the towns and villages. Kansas has a belt of limestone, suitable for building, that stretches across the eastern-central portion of the State from the valleys of the Blue and Nemaha to the Arkansas. These stones are compact in their texture and of different colors, such as light gray and creamy buff and gray-buff. "The most extensive quarries in the State are in this belt, in the vicinity of the villages of Strong and Florence. This stone is of such texture that it is easily carved into designs, and fashioned into all shapes employed in architecture. The State-House at Topeka is constructed of this stone." (*Report of State Board, vol. ix, pp. 501, 502.*) Limestone occurs in many other localities in the State. Nebraska is also well supplied with different grades of limestone.

Montana has an abundance of limestone that can be used as building material and for other purposes, as well as excellent marbles. Wyoming is credited, in the Governor's report (1885), with being very rich in limestone in "all sections of the Territory. It [the stone] is of a tenacious, hard, metallic quality, and susceptible of marble-like polish; it is used extensively for building as well as for other purposes."

Colorado has an abundance of limestone that is found to be excellent building material, the quarries of which in Jefferson County are deemed practically inexhaustible. From this natural deposit are obtained blocks of stone, varying a little in color, but becoming hard and durable when exposed to the air. On the Pacific slope, California, Oregon, and Washington Territory all have limestone in abundance, but of a quality that is not very available for building purposes.

Sandstone.—As a building material sandstone is the most common that Nature has provided; it is more widely diffused than any other, and has more marked diversities in respect to its texture or compactness, as well as to its various shades of color. This stone usually occurs in deposits that are moderately large, but so numerous and scattered as to be easily accessible to supply the wants of the people.

Brown-stone.—East of the Alleghanies are located two very important deposits of sandstone that is much used in constructing buildings—one in Connecticut, in the valley of the Connecticut River; the other in New Jersey, in the vicinity of Belleville.

In these deposits exist one or two grades, in respect to the texture of the stone, some being less compact than others; as to its appearance, the predominant color is a reddish brown, while the subordinate shade is lighter, but both retain their natural color when exposed to the climate. These beds of brown. stone are immense, and for more than a third of a century have furnished an enormous amount for buildings in the cities and villages within reach.

Blue-stone.—In Ulster County, New York, are extensive beds of a very hard blue-stone, that is much used for adjuncts to buildings, as in steps, lin. tels, sills, etc., while it is in great demand in the cit. ies for flagging and causeways in the streets. New York has also an abundance of sandstone located at Potsdam, very hard in its texture and durable, of a red color interspersed with grayish veins, but of a grit almost as hard as granite.

As we pass south along the Atlantic slope, we find almost an unlimited number of comparatively small beds of sandstone, but sufficient to supply the local demand. East Tennessee, however, has large depos. its of a light-colored sandstone amid the Cumberland Mountains, while an isolated range—for some un. known reason called Niagara—furnishes an abun. dance, of an excellent brown-stone; also in the west. ern portion of the State is found a peculiar building material designated *iron-sandstone*, which is used ex. tensively for foundations.

Sandstones of the Valley.—The States within the

valley of the Mississippi are remarkably well provided with sandstone for building purposes. These
stones are of nearly every grade in respect to texture, some being very fine and almost capable of receiving a polish; the same may be said of the numerous varieties of sandstones amid the Rocky
Mountains, and of the Pacific coast. Their colors
are also various, consisting of different shades of
drab and of blue, and of a blue tinted with gray;
and pure gray, and with a greenish tinge; others
white, or red, while some are quite dark. Vast
quantities of these sandstones find their way to cities
and villages near home for buildings, and also east
of the Alleghanies, sometimes for houses, but much
more for use as trimmings of windows, etc., in brick
edifices.

The Amount of Sandstone used.—The deposits
of this building material are so numerous and so
widely distributed throughout the Union, that we
can estimate their area and depth only by taking as
a criterion the amount quarried and used. The
number of cubic feet of sandstone thus used in one
year, as is learned from the census of 1880, is in
round numbers 25,000,000. Of these Ohio is credited with one third, Pennsylvania with one fourth,
New York one eighth, New Jersey one twelfth, and
Connecticut one twenty-fifth, while the remainder is
divided between twelve States and one Territory.
During the same year, 1880, the number of cubic
feet of marble used for building, and limestone for

that and numerous other purposes, was 66,000,000. Of these Illinois produced one fifth, Iowa and Pennsylvania each one sixth, and Missouri one sixteenth; the remainder being divided between fourteen States. Likewise in the same year were produced 21,000,000 cubic feet of "crystalline silicious rocks" or granites, of which Massachusetts produced one fourth, Connecticut and Pennsylvania each one seventh, Maine and New Hampshire each one tenth, Rhode Island, Maryland, and Virginia each one twentieth, while the remainder is divided between ten States. To these vast amounts of building-stones thus recorded by the census are also to be added the immense quantities that the people use from their own quarries, and of which no record is ever published.

SLATES.

The chief quarries of slate in the Union are found in the States of Pennsylvania, Vermont, Maine, and New York, and in the order named. Pennsylvania, having very large deposits to draw from, produces more slate than all the other States of the Union combined. Her most extensive quarries are in the vicinity of Allentown, on the Lehigh; they furnish a slate of a fine texture, generally darkish gray in color, and specially adapted for use in schools, for roofing, etc., as well as for the almost innumerable purposes to which slate has been applied within recent years.

The slate deposits of Vermont and New York

20

belong virtually to the same slate area, as they are near each other, the State line merely running across them. Castleton, in Vermont, and Middle Granville, in New York, may be deemed important centers in their respective districts. The slate of both areas partakes of the same general and excellent qualities, comparing favorably with that of Wales, both in fineness of texture and in color, while it is found to be adapted to all the purposes to which slate is ordinarily applied. It is also susceptible of being enameled, and made to imitate the best variegated marbles, and of any pattern that the taste of the workman designs—the colors being baked in effectually. This marbleized material is extensively used for mantels, wash-stands, for tops of tables, and other furniture where it is applicable. It is very durable, and not liable to be injured under ordinary circumstances. The large deposits in Vermont, and the adjoining ones in New York, are in a range of hills which extend north and south for more than twenty miles. In Vermont are also very large deposits of a different grade of slate, one that is of a light drab color, soft and of a delicate texture and free from grit, from which, in untold millions, ingenious machinery makes slate-pencils.

In New England, Maine and Massachusetts have each large deposits of slate of good quality, while Maryland, Virginia, and Georgia have large beds of slate that is suitable for the purpose to which it is usually applied.

XXVIII.

THE State of Vermont furnishes us the greater portion of this fine class of marbles. A remarkable belt of territory, which contains numerous isolated beds of marble, commences in Westchester County, New York, and, with a short breakage, extends northeasterly across Vermont into Canada. The southern end of this belt, as already noticed, furnishes a coarse marble for building purposes; but in passing northward the quality of the marble improves, its texture becoming more fine and the color a purer white. Within about thirty miles north and south of Rutland the finest marble of this belt is found. The veins of the best marble are nowhere very large, and the finest quality lies the deepest; the dip of the vein is sometimes forty-five degrees. The more pure and delicate the marble is, the more easily is it injured by the jars occasioned by quarrying, and in consequence a valuable vein is sometimes cracked and ruined for quite a distance. This difficulty has been mitigated by means of ingeniously constructed machinery which drives saws of a pecul-

iar character—often with teeth of black diamonds—
that cut channels around the block designed to be
taken out. These deposits appear to be practically
inexhaustible. The marble is of a delicate texture,
and the color white, some classes being suitable for
statuary; it all receives a fine polish, and furnishes
beautiful and useful ornamentation for furniture and
for interior decorations generally.

Southern Marble.—The deposits of marble in the
southern end of the Alleghanies have attracted much
attention, not only because of their immense size, but
of the characteristics of the marbles themselves. Instead of being white alone, as in Vermont, they are of
almost every shade of color, but of equal fineness in
texture. These deposits are found in different parts
of three States. In the western portion of North
Carolina are beds of excellent marble, the latter being of various colors, such as, "white, pink or flesh-
colored, black, gray, drab, and mottled." These
marbles are susceptible of being finely polished, and
are very valuable for ornamentation, and for interior
work in buildings.

Serpentine is a mineral stone usually of an obscure green color, with shades and spots that give it
a mottled appearance resembling a serpent's skin.
In the same section of the State occurs a very large
deposit of serpentine or verd-antique—the common
name being green marble, though strictly it does not
belong to the class of marbles, it being chiefly a silicate of magnesia. This material is found in Wake

County, in such large blocks that it is "sawn into slabs and polished and sold under the name of green marble." Remarkably beautiful specimens are ob-tained from quarries some dozen miles from Raleigh, North Carolina, and also in the vicinity of Patterson, Caldwell County. The latter has a dark greenish-black color with fine veins of yellowish green; in another bed, in the same county, the serpentine is greenish gray in color, while in a deposit, in the neighborhood of Asheville, the color is dark green. All these grades of serpentine receive a very fine polish, and are used much for interior decorations. No other large deposit of this mineral stone occurs in the Union.

East Tennessee is peculiarly fortunate in the beauty of her marbles, as well as in the extent of their deposits. The pure white does not exist in this State, but instead "a remarkable commingling of one or more brighter colors with the pearly tints of the sea-shell." Nearly all the Tennessee marbles belong to the variegated class; in some localities the foundation color is drab or dove; in others a gray or pinkish-gray, and, in others still, of a mottled pink and gray—pronounced to be as beautiful as the famed variegated marbles of Italy. These mottled colors are fully displayed by means of the fine polish which the marble is capable of receiving. In elabo-rately finished private dwellings, in the cities on the Atlantic slope, these marbles are often used for in-terior work of different styles, but more especially for steps and wainscoting of stairways, in hotels and large and finely finished apartment-houses, as in the city of New York. These marbles are used in the interior decorations of the Capitol at Washington, and in the Governor's room in the new Capitol at Albany, New York.

Georgia has a marble belt that extends through a number of counties, and which by far is the largest yet discovered in the Union. These marbles are of uniform fine texture, but of different shades of color; some "white statuary marble and several variegated kinds, some of which are unique in color and remarkably beautiful." In one district is a red variegated marble, as well as that which is cream, flesh, and dove-colored; in another, "the marble when polished presents the beautiful effect of a network of white lines on a dark blue or black ground." Upon the whole, the marble area of these three States is beyond compare the most extensive in the Union, and the most diversified in character. (*Handbooks of North Carolina and Tennessee, and Commonwealth of Georgia.*)

Iowa and Missouri Marbles.—Iowa has a unique stone known as "coral marble," thus named from one of its constituents, which is a "coral-like fossil" or "fossil-sponge," but the stone is sufficiently compact to take a perfect polish. "It is a trifle harder than Italian marble, and is remarkably free from the checks, seams, and defects common to most colored marbles." "Though no two pieces are perfectly alike, the groundwork of the color is mostly buff, gray, or drab, and this is inlaid or blotched with masses of coral varying from one to twenty inches in diameter and of the most exquisite and delicate coloring and tracing." Of the latter some resemble wood, others sea-shells; some having specks of pure white, others veined in a dark mahogany-brown. The beds of this marble are said to be practically unlimited in extent; they are located in the northern portion of the State, in the valley of Cedar River, and in the vicinity of Charles City. (*Res. of Iowa, p. 73.*)

Missouri has several and quite extensive deposits of excellent marble, some of which is fine-grained and durable, and also has beautiful shades of color. " One of these varieties is commercially known as onyx or onyx-marble, a stalagmite formation found in beds of caves," and said to occur nowhere else in the United States. It resists acids and does not stain; it is extensively used for mantels, fine furniture, etc. This State has vast quantities of red marble mixed with flint, which gives it a variegated appearance; and also a large deposit of white marble, with blue streaks that gives it a beautiful luster when polished. (*Missouri Handbook, pp. 29, 187.*)

Other Marbles.—The Territory of Montana has beds of excellent marble, and Wyoming has a large deposit on the Laramie Plains, the marble of which is " crystalline in character, very fine-grained, and yields a high and beautiful polish." Marble is also found in other localities within the Territory.

Marbles occur in New Mexico and Arizona, but as yet the deposits are undeveloped. Colorado, among its other natural wealth, has marbles in abundance. They are of a white color, slightly seamed or streaked, and of good quality. Oregon has rich deposits of marble, and so has Washington Territory. The latter are but partially developed, as the wants of the people have not yet demanded their use to much extent.

XXIX.

ABRASIVE MATERIALS.

Corundum and Emery.—Corundum is defined as "the earth alumina in a native crystalline state," nearly pure; while emery, having the same alumina as a basis, contains from one fifth to one third of iron oxide. The latter occurs sometimes in nature in masses and grains, but is oftener prepared for use. The particles constituting corundum, or emery, are exceedingly hard, in that respect almost equaling the diamond, so that, after being reduced to the proper fineness, it is employed in abrasing or wearing away surfaces by means of friction, and thus "it is used in the arts, for polishing metal, hard stones and glass." It has already been noted (p. 262) that, in veins or beds of corundum, the alchemy of Nature often produces one class of gems—the beautiful sapphire, with its clear and brilliant colors of blue, red, and purple, emeralds, and topazes.

The coarser variety of corundum, so useful for certain mechanical purposes, occurs in a very large number of localities—more than fifty are already known—extending from Massachusetts to Alabama, while "emery has not been found in any great quantity within the limits of the United States." At Chester in the former State is found a deposit of corundum so large and so accessible that parties have been induced to prepare the raw material for use in the arts; but at present the original bed is fast becoming exhausted, and the corundum is now mostly supplied to the works from other sources, especially from North Carolina; the very small and isolated deposits in the intervening States yielding little or none. A belt of territory 100 miles wide, and in which corundum occurs in isolated deposits, extends from the Virginia line in a southwest direction across the western portions of North Carolina, South Carolina, and through Georgia to Dudleyville, Alabama. The most prominent deposit is at Corundum Hill, Macon County, North Carolina. The mineral is found in "pockets" or detached places amid the rocks; sometimes it occurs in large masses, but mixed with extraneous substances, then in crystalline forms. In one instance a crystal was found that weighed 312 pounds; the latter is in the cabinet of Amherst College, Massachusetts. (*Handbook of North Carolina, p. 180.*)

It is stated that a large deposit of emery has been discovered in the vicinity of Peekskill, Westchester County, New York; also it is found near Allentown, Pennsylvania, and in the State of Delaware, and in Pettis County, Missouri.

Buhr-stones.—The buhr-stone is composed of a

flinty quartz filled with minute cells which have sharp edges or grit, a property that fits the stone for grinding grain. These stones have hitherto been imported, but at present stones having similar qualities have been discovered in a number of places in the United States, and have been extensively utilized. In respect to grinding grain, the time-honored millstone is being superseded very rapidly by the system of using iron rollers in converting wheat into flour. The native buhr-stones, for the most part, are now used for grinding the coarser grains, as corn, and in making oatmeal, in pearling barley, in grinding paints, chemicals, fertilizers, etc.

This stone occurs in many localities, of which a prominent one is in Ulster County, New York. It is known as the Esopus stone—a granular quartz, but of variable texture and hardness; also in Lancaster County, Pennsylvania, where the stone is found, not in the usual beds or quarries, but in the form of bowlders scattered over the surface. In North Carolina is found a silicified " shell-rock " forming one grade of buhr-stone ; the latter have supplied almost all the domestic demand ; and in Georgia this stone appears in large quantities in one place on the banks of the Savannah River. " It varies from a light gray to a reddish brown color, and abounds in cavities lined with chalcedony." Buhr-stones of various grades of excellence have been discovered in limited quantities in the States of Alabama, Arkansas, and Missouri, and they are also reported to exist in Pitt River County, California.

The State of Ohio has large deposits of a sandstone, which is ordinarily fine-grained, of great hardness, and with a very fine grit. These beds are in the valley of the Cuyahoga River, near the village of Peninsula, and the stones obtained here are remarkably well adapted in their texture for milling oats and barley.

Grindstones.—The belt of territory, in Ohio, that furnishes stones of a texture essential for good grindstones extends south for about 140 miles with varied width, from Berea, within 20 miles of Lake Erie, to Marietta, on the Ohio River. There are within this belt, following the valley of the Cuyahoga and over the divide into that of the Muskingum, seven very large deposits of this grade of sandstone. In each of these quarries the texture and the color of the stone are somewhat different from those of the others, though all are included under the general term the " Berea grit." At Berea, the most prominent deposit, the stone is white in color, and has a fine sharp grit; the next, toward the south, is brownish white, with a soft loose grit; the next, grayish white, with a coarse sharp grit; the next, yellowish in color, also with a coarse sharp grit; while the quarry farthest south, in the vicinity of Marietta, produces a comparatively soft stone, the grit of which is very coarse, but is adapted for heavy work. The grindstones made here are very large, reaching as high as twelve inches in thickness and seven feet or more in diameter. Extensive manufactories of these classes of grindstones are located in the city of Cleveland.

Michigan Stone.—It will be noticed that, proceeding from the south toward the north, the general texture of these stones becomes finer. About 200 miles directly north of Berea, in the Lower Peninsula of the State of Michigan, on the extreme northeast point of what is called the "Thumb," jutting on Lake Huron, occur immense and practically inexhaustible deposits of another class of stone from which also grindstones are made. (*Statistics of Michigan, 1882, p. 207.*) This is a remarkably fine-grained, argillaceous (clayey) stone, free from foreign substances, and of a uniform texture and blue color. It is deemed perfect in its adaptation for finishing, where a very fine and delicate edge is required. It would seem as if these stones of Ohio and Michigan were adapted to supplement each other—the one doing the coarser and the other the finer work. The grindstones produced in the former State are much larger than those made in the latter. In Nova Scotia is the only other place yet discovered on the continent where are found materials suitable for making grindstones of a high grade. There is, however, scarcely a State in the Union but which has more or less of certain grades of stones that have been used in domestic work as grindstones, though their work must be reckoned very inferior, when compared with that which is done by those from the great quarries.

Whetstones.—*Novaculite* is the scientific name given to a variety of silicious rocks of different shades of color from dark to honey-yellow, from which hones are made, for securing the finest sharpening of surgical instruments, razors, etc. The common name is oil-stone or oil-whetstone, as that substance is applied when the stone is used. A large deposit of this material occurs in North Carolina, in the vicinity of Chapel Hill. This has been worked to supply the wants of the people in that section of the State. The stone is also found in other localities in that region, and also in Oglethorpe and Lincoln Counties, Georgia, there being immense beds near **Graves's** Mountain; this has also been utilized to some extent for domestic purposes.

As far as discovered, the most important deposits of this oilstone in the Union are in Garland and Hot Springs Counties, Arkansas—in the latter at Whetstone Mountain. This extensive deposit has supplied more of these stones for use than perhaps all others combined, the district whence the oilstones are taken being about fifty miles long by twenty in width. While a portion of the stones are dressed at the quarries, the much greater part are sent elsewhere to be finished for market. New Albany, Indiana, has an establishment for manufacturing the stone in all classes and grades, to supply the demands of the country. Oilstone is also found in Orange County, Indiana. It is reported as of good quality. The harder varieties of this fine-grained stone are also used by engravers.

Lithographic Stone.—Prof. A. R. Roessler (geologist) discovered recently in the valley of the Colorado in Burnet and San Saba Counties, Texas, an extensive deposit of lithographic stone that has all the excellent qualities of the famed Bavarian stone, Germany, which has been used hitherto in the United States exclusively for fine work. " This stone presents all the appearance and physical characteristics of the best lithographic stone of Bavaria. . . . It is yellowish-gray in color; has no fibers, veins, nor spots ; a steel point makes an impression on it with difficulty." (*Prof E. Everhart, chemist, State University, Texas, 1885.*) " The Texas stone in its chemical composition does not differ materially from the Bavarian, and wherein it does in silicates, it has the advantage." It has been " tested practically under a press and pronounced equal to the Bavarian," by a lithographic company in Chicago. (*Galveston News, April, 1885.*)

XXX.

GRAPHITE, OR PLUMBAGO.

THIS is a mineral composed of carbon almost pure, for in it there is often a trace of iron; it has a metallic luster, and leaves a dark, lead-colored trace upon paper—hence its original and common name, black-lead. Though now theoretically considered as derived from vegetable tissue, its mode of origin has by no means been accounted for satisfactorily; and Nature, though often interviewed by chemistry, has hitherto given out only vague hints, but in the main has persistently refused to reveal the process by which she distilled this unique mineral.

Uses to which applied.—Graphite, or plumbago, is utilized in very many ways; as in the manufacture of pencils—which requires its finest quality—from the lowest grade used by the ordinary workman to the highest demanded by the artist; also in *dry lubrication*, as in producing a smooth and easy action in the piano and the organ, or in any wooden surfaces that move over one another. To fit the plumbago for the latter use requires care in removing grit.

Within recent years the application of the coarser

grades of graphite in mechanical industry has increased in an unprecedented manner. By being properly mixed with an oily substance it makes a perfect lubricant for machinery of all grades, removing the friction in a remarkable manner, for as a lubricant it is not affected by heat or cold, nor by steam nor acids, and it readily coats the surfaces, passing over one another with a shiny sort of veneer. Still more remarkable is the utilizing of that property by which it can not be melted by heat produced in the industries; hence its demand in the form of retorts and crucibles—the latter especially, in the mode of converting iron into steel by the Bessemer process. It. is also used for pigments and for polishing iron surfaces, and in foundries in making facings for the molds in order to produce a smooth surface on the vessel cast. " This has been a great improvement in American foundry practice."

Where Graphite is found. — This mineral is widely distributed in the United States, but thus far in only one or two localities does it exist in sufficient quantities and purity to warrant its mining, as it is so often mixed in the bed or vein to a large extent with slaty rubbish and other foreign substances. " It is found in the rocks of the Alleghanies from Alabama to Canada, but in no instance thus far has a deposit been discovered of commercial value." It occurs in different places, but in small quantities, in North Carolina, the beds in Wake County being the most important. The belt of territory in which

21

it occurs extends for some seventeen miles, passing near Raleigh. Some of these beds have been
opened to supply the domestic demand, but "the
deposits are more or less impure, the graphite being
of a slaty and earthy variety." In Georgia, a mine
in Elbert County is the largest deposit yet discovered in the State; there also occurs graphite in
small "pockets" amid the rock in a few localities in
Northwestern South Carolina; this is reported to be
excellent in quality. "The deposits [of that whole
region] are of such a nature that purification is
economically impossible." (*Min. Res. of U. S., 1882,
p. 590.*)

In California, near Sonora, Tuolumne County, is
a large deposit of graphite, but it has been impossible to obtain it sufficiently pure from the bed for
commercial purposes, as the original material is so
mixed with slate and other foreign substances as to
make the separation of the pure graphite too expensive. The latter is found in small quantities in several localities in the State and also in Nevada, but
"these possess just now no special value." Graphite
is also reported to be in abundance in Wyoming,
and in deposits in the Black Hills, in Dakota.

New England and New York Graphite.—Graph-
ite occurs in very limited quantities, in Hinsdale,
North Brookfield, and Sturbridge, Massachusetts; at
Brandon, Vermont; and Ridgebury, Connecticut.
The only place in the Union where graphite is now
mined successfully is at Ticonderoga, Essex County,

New York. "The vein there has been worked to a depth of 600 feet. . . . The graphite schist or vein is about fifteen feet thick, carrying from 8 to 15 per cent of graphite, and is practically inexhaustible. This mine is unusually rich, for it perhaps produces the finest graphite in the world." (*Min. Res. of U. S., 1883–1884, p. 915.*) It also has a coarse grade for making crucibles and a very fine one for making pencils. These are separated by thin partitions in the rock; the fine, being of smaller quantity, is in isolated cavities called "pockets."

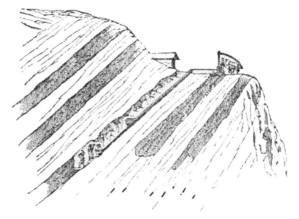

FIG. 12.—Section of Graphite Mine at Ticonderoga, New York.

Florida Cedar.—We may here remark that Nature has given us the monopoly of the world in the peculiarly fine-grained cedar for making pencils. This wood is only found in certain swamps in Florida, and it is used for that purpose both in this country and in Europe.

XXXI.

THIS substance, so essential for man's health and comfort, is produced in great abundance in many and widely diffused localities within the United States—from New York to California, and from Michigan to Louisiana. Chemically speaking, salt is the chloride of sodium; the latter is "a yellowish-white metallic element, soft like wax and lighter than water." The proportions are sixty parts of chlorine, by weight, to forty of sodium. Nearly all common salt is obtained by means of evaporation, a process in which, by either boiling or by the application of solar heat, the water of the brine is made to pass off in vapor, while the saline particles are left in the vessel. In the main there are two classes of brine, or salt-water—that of the ocean and that of springs or wells.

Salt (where found).—East of the Alleghanies, New York State is the great producer of salt, which is derived principally in the middle part of the State from what is known as the Onondaga salt-bearing district, the city of Syracuse being deemed the center. This is an area immensely rich in brine, or salt-water, both in quality and in quantity; indeed, it appears that the middle portion of the State is underlaid by salt in some form. Within recent years a discovery of rock-salt was made accidentally near the village of Warsaw, Wyoming County. An artesian well was being drilled for oil, when, at the depth of about 1,500 feet, the drill encountered and then passed through a bed of rock-salt forty-six feet thick; soon after a similar discovery was made about six miles from the former. Near Ithaca, south of Syracuse, another bed of rock-salt, 250 feet thick, was found (1887) at the depth of 2,600 feet, while sinking an artesian well for natural gas.

Onondaga District.—From the earliest times salt was obtained in the Onondaga district, and the operations there carried on are to-day on an immense scale. The salt-water is drawn from wells almost innumerable, and is run into large vats or shallow reservoirs, so as to expose as great a surface as possible to the heat of the sun; these vats number more than 40,000, while there are also engaged several hundred factories in boiling the brine, and preparing the salt for use. The finest grades made here are deemed as pure as the famous Ashton salt produced at the mines of Cheshire, England. Some years since the National Government subjected this salt to a very severe but impartial test, in competition with that of Turk's Island. The test was carried out in preparing provisions for the army and the navy.

The Syracuse salt proved to be the more effective, purer, and better in its effects upon the meats thus prepared. In the Warsaw district operations in making and refining salt are on a large scale. Rock-salt and brines are also found at Saltville, Smythe County, Virginia, and in one or two other counties, but the production is comparatively not large. There are a few salt-wells in North Carolina, but the output is quite limited.

West Virginia and Ohio Salt.—The former State has a number of localities in which brine is obtained; one on the Kanawha River—of which district the village of Charleston is the center, and where salt was first manufactured, even when the Old Dominion held sway in that section—these works have been prosperous. Tradition tells in what way the hidden treasure was discovered. Two boys, when the water in the river was very low, went fishing, and took with them a bottle of milk as part of their lunch. When returning home, they noticed a spring of pure sparkling water bubbling up close to the edge of the river's stream, now very much reduced in volume, and being thirsty they filled their bottle from the spring, and were astonished to find it *salt-water*. The news of the incident spread; experiments were made by boring, and soon brine was found in abundance, and which has been flowing from that day to this. Salt-water has been discovered in many other localities within the State, among which is that in Mason County, in the vicinity of West Columbia, on the banks of the Ohio. Upon the whole, this State is rich in the abundance of its brine, and in the necessary coal-fuel to operate the wells, but the brine itself is not as rich in saline properties as are those of New York and Michigan. There are four different qualities of salt produced in West Virginia, which are characterized as coarse, common, fine, and dairy.

In the southeastern portion of the State of Ohio, and near the Ohio River, is a prominent salt-producing district, being similar in many respects to that of West Virginia.

Michigan Salt.—In this State what is known as the salt-bearing district is confined almost entirely to parts of some half-dozen counties lying on or near Saginaw Bay. Here is produced an immense amount of salt, and the reservoirs in the earth from which the brine is pumped are apparently inexhaustible. These artesian wells are near or on the bay, and in the short valleys extending into it. "The salt group of the State has a wide extent, though thus far the salt industry has been confined exclusively to the Saginaw Valley. . . . The reservoirs of brine lie at a depth reaching to more than a thousand feet below the surface of Lake Michigan." There are about 260 salt-wells in the State; of these the average depth in the Saginaw Basin is 950 feet, though some wells run to 1,900 feet. (*Commissioner's Report to the Governor of Michigan, 1882, p. 32.*)

Salt in the **Great** Valley.—Saline springs and salt-wells are quite common in some of the States in the Great Valley, as in Illinois and Indiana, across the Ohio River in Kentucky, and Tennessee, and beyond the Mississippi in Nebraska, Missouri, Kansas, and Arkansas. The latter has deposits of rock-salt in Dallas and in Hot Springs Counties, though thus far they have been little worked. Kansas has, in the valley of the Solomon River, saline springs of great value. The brine flowing from the latter has been tested and found to yield a large percentage of salt.

Louisiana Salt.—This State has two widely separated localities where salt is found: one in the northwestern portion, known as the " Licks "—thus named because the forest animals resorted to them to lick the salt; the other is an enormous mass of rock-salt, lying in almost a due south direction from the " Licks," and within a few miles of the Gulf, on an island "in a sea-marsh" near New Iberia. In the " Licks" are numerous salt springs flowing into the adjacent streams amid the salt-marsh sedges. The water, which is not very strong—only two to three per cent—in saline properties, is obtained chiefly from pits or shallow wells dug in the flats. On the other hand, the mass of rock-salt on Petite Anse or Avery's Island, is remarkable for its dimensions, and the purity of the article. The mine is opened by a shaft, from which "the mass is worked by a system of chambers and cross-headings, thirty-five to forty-two feet wide by sixty-five in height, leaving a roof of fifty-five to sixty feet of solid salt above, and pillars of the same diameter as the chambers (forty-two feet square) for its support." The shaft is sunk to the depth of 190 feet and through 165 feet of solid salt, which is unchanged in character. The area covered by this mass is 144 acres, and the amount of hard salt is estimated by a competent engineer at 28,600,000 net tons!

Rocky Mountain Salt.—In the form of salt springs, indications of hidden stores of brine occur in these mountains, extending from Montana to New Mexico. In the Yellowstone Valley such springs are numerous, while in Wyoming rock-salt of great purity is reported as abounding in extensive beds. In Dakota are also a number of springs that furnish excellent salt and for which the flow is abundant. South Park, in Colorado, contains numerous saline springs which have been utilized sufficiently for local demand. New Mexico has many lakes that furnish large quantities of good salt. Socorro County obtains common salt from lakes; and in Valencia County, on the plateau between the Rio Pecos and the Rio Grande, are numerous lakes, where large deposits of excellent white salt are found, and which can be had free for the collecting and carting. (*Illustrated New Mexico, p. 232.*) Salt lakes occur also between the Sacramento and Organ Mountains; there is also a large one sixty miles south of Santa Fé. From these and similar ones, where salt is left by the natural evaporation of the water, the greater

portion of the Territory derives its supplies of that essential article.

Texas Salt.—This State has extensive lagoons, or sea-marshes, along the coast in the vicinity of Corpus Christi, where salt is deposited as the result of solar evaporation, as the very high tides leave a portion of their water to be thus acted upon. Numbers of small saline lakes are found in different places in the interior of the State ; these supply more or less the local demand, but, since the advent of railways, salt is brought in at such reasonable rates that what is derived from the lakes is now quite limited. Near El Paso are several salt lakes that are comparatively large.

Utah and Arizona Salt.—The waters of Salt Lake supply sufficient salt to meet the local demand, and one hundred miles south from the city, near Nephi, is a large deposit of rock-salt; while in the southern portion of the Territory are other extensive deposits of the same kind, of excellent quality. " This resource of the Territory is destined to become very important, as the supply is practically unlimited." In Eastern Arizona is a shallow lake, in which salt is precipitated on the bottom, by means of the heat of the sun evaporating at one season the water that accumulates at another. Here the inhabitants, farmers, and stockmen, help themselves from this common property, by coming at certain times with wagons, and shoveling up the salt, and carry away as much as they need.

Oregon **and** Idaho Salt. — Saline springs are abundant in Oregon, and in their vicinity reservoirs of brine seldom fail of being found by boring. In Southern Idaho, in Oneida County, are numbers of saline springs from the waters of which hundreds of tons of salt are annually obtained by means of solar evaporation, even more than is required for local domestic use, and for smelting purposes.

Pacific Slope Salt.—This section appears to be abundantly supplied with facilities for obtaining this mineral. Here are saline springs, and ponds, and lakes containing water from these springs, and sometimes great mountain-masses of crystalline layers of rock-salt, and deposits in the beds of lakes long since dry or nearly so.

Nevada Rock-Salt.—Of this the largest deposits on that slope exist in this State. In its extreme southeastern portion, in Lincoln County, a few miles north of the Grand Cañon, in the Colorado 'River, and on the river Virgen, a tributary of the latter, is a formation of rock-salt resting on granite, and so extensive as to constitute a notable portion of the mountain itself. This formation as a bluff extends along the eastern bank of the Virgen for twenty-five miles or more, and in some places it is several hundred feet in height. More than 60 per cent of the entire mountain or cliff consists of hard rock-salt, which itself contains 90 per cent of common salt, having the pale-green color and transparency of ice. Twenty miles farther up the river, on its west side, is another hill of rock-salt, but less extensive; the color of the mineral *here* is of dazzling whiteness. These rich and extensive deposits are in trust for future generations, as at present transportation is afforded by neither river nor railway.

Nevada has also saline springs in the northern portion of the State, near the line of the Central Pacific Railway, in Churchill County. Here salt is manufactured chiefly by solar evaporation, and the silver-smelting establishments at Virginia City, and in the vicinity, are supplied principally from these springs and wells. About eighty miles east of the latter city is a remarkable and extensive saline marsh in a depression near Sand Springs. "There is spread over the marsh an incrustation of impure salt from two to three inches thick, brought up by efflorescence from below." When this surface incrustation is removed, another begins to form; the process goes on so rapidly that several crops can be collected in a year. This impure material can be used in metallurgical operations, such as in obtaining silver from the ore. There are many other localities where salt has been discovered in the State, but have as yet been but little utilized, such as in Esmeralda County, where salt is found in connection with immense alkali flats. In one instance "hundreds of acres are underlaid by a hard-pan of solid salt, more or less mixed with mud and sand." (*Min. Res. of U. S., 1882, p. 545.*)

California Salt.—This State has abundant facilities for supplying hersell with salt, not only by solar evaporation from sea-water, in lagoons, but also from saline springs; and within her borders more is obtained by means of the former process than otherwise; the salt is afterward purified for domestic use. " The salt made on the Bay of San Francisco is as good as any in the world, being in strength and purity equal to the best French brands, and much heavier than the imported from England." (*Min. Res. of U. S., 1882, p. 549.*) This State has salt marshes nearly as extensive as those in Nevada; numbers of these are waiting to be utilized. The Bay of San Francisco alone, with its lagoons, furnishes not less than 25,000 tons annually. The amount of salt used in that section of the country is exceptionally large, as in the reduction of silver-ores, and in other such operations, about 30,000 tons are consumed each year: to this must be added what is consumed in domestic use.

Upon the whole, as we have seen, the United States are amply supplied with this invaluable article from native springs and from wells, while in addition, il need be, are afforded numerous facilities, especially along our Southern coasts, to obtain it from sea-water by solar evaporation.

BROMINE.

Bromine is found, but in only very small quantities, in the water of the ocean and in saline springs and wells. In the waters of the great salt-wells of

New York and Michigan it has scarcely a trace, while in those of the adjacent salt districts of West Virginia and Ohio it occurs in quantities sufficient to warrant its manufacture, and only here in the Union is it made; yet the United States holds the first rank in the world in its production.

"Bromine is a deep, reddish-brown liquid, of a very disagreeable odor." It is used medicinally, in photography, and in analytical and experimental chemistry. Sometimes it is used as a disinfectant and for bleaching purposes, and also occasionally in producing colors. After the salt has been extracted, some waters leave as a refuse a liquid sediment known as "bittern," an ingredient of which is bromine. From this substance it is extracted by an elaborate chemical process. Bromine is usually found to have a trace or slight connection with chlorine and iodine. The latter is not produced in the United States, but is obtained for the greater part from ashes of "kelp," the result of burning certain classes of sea-weeds found on the shores of Ireland, Scotland, and Norway. The purposes to which bromine is applied are quite limited, and in each of these the quantity is not large. In 1885 the product in the United States was about 310,000 pounds, and that for the most part supplied the world.

BORAX.

Borax, a substance usually of crystalline form, is of a white color, or sometimes grayish, and often with a shade of blue or green. Being an excellent flux, it is used in many metallurgical operations, and in ordinary delicate soldering. " The leading uses of borax are in welding (for which the greater part is consumed in iron and steel manufacturing); in refining metals as a crucible flux; in enameling; by packers, in preserving meats; and, because of its cleansing power, as a detergent for household purposes." (*Min. Res. of U. S., 1882, p. 576.*)

Where obtained.—This valuable substance or salt is scarcely found in the Union except in the States of California and Nevada, though in some of the hot springs in the Yellowstone Valley, Montana, boracic acid occasionally appears. An enterprising California physician—Dr. John A. Veatch—detected the presence of boracic acid in certain mineral springs of the State; he never relaxed his efforts till, after much search and travel, and many disappointments, he discovered a large deposit of borax itself, resting in its bed on the margin of a marsh. It was in semi-opaque crystals imbedded in the mud in the marshy soil—some of the crystals very small, and others quite large. This was in Lake County, and on the border of Clear Lake. This deposit was worked, and meanwhile borax was found in several other localities within the State, such as in the Slate Range marsh in San Bernardino County. This deposit is of rather more than average richness. Additional deposits were afterward found in the same county. A number of other borax-bearing districts occur in Inyo County; two of these contain several thousand acres, and appear to comprise the chief deposits in that region; and there are still others, of minor importance, in the State.

The borax-bearing districts of California are small compared with the immense ones found in Nevada. In the latter State are a number of extensive marshes in which the boracic salts predominate, and they appear to be practically unlimited; these salts are the result of the uniting of boracic acid in nature, principally with lime or soda. In Nevada borax "crystallizes in long, silky fibers, which gather into balls from an eighth of an inch to two or three inches in diameter. These globular masses have the luster of white satin." In some of the marshes or alkali flats of the State the different combinations of borax produce salts, the efflorescence of which, at certain seasons, "is moist, flaky, and of dazzling whiteness, and might easily be taken for fallen snow." (*Min. Res. of U. S., 1882, p. 568.*) "In the barren wastes, where nothing grows and everything looks desolate, is found the wealth that lies in the vast fields of borax, salt, niter, soda, sulphur, etc. . . . The borax districts of Esmeralda County are proving to be of great value and extent." (*Report of Surveyor-General to the Governor, 1885.*) These four prominent marshes or basins, to which allusion is here made, cover from 10,000 to 20,000 acres each, and, all combined, about 115 square miles. At quite separate points in the State are an additional number of borax marshes or districts, so that Nevada is to be pre-eminently our producer of the borax of commerce, the chief market of which is in New York and London, while Germany, with China and Japan, come in for a share. The borax produced (1885) on the Pacific slope amounted to 4,000 tons.

Saltpeter, or Niter. — This substance occurs in nature as a crust of minute silky crystals. It is used as an antiseptic, in making gunpowder, and in medicine. Though often found in a native state, before using it requires refining. In this native form it occurs in numerous places in the United States, especially in the western portion, such as in Utah and Nevada, but the greater amount used at present in the Union comes from Peru and Chili. Deposits of the nitrate of soda, similar to that in Chili, have been found in Humboldt County, Nevada, within twenty-five miles of the Pacific Railway. Its crystals either occupy crevices in the rocks, or are imbedded in the earth from a few inches to thirty feet beneath the surface. The geological and climatic conditions on the Nevada " Forty-mile Desert" are similar to those in Peru and Chili, where this nitrate of soda covers the dry desert for several feet thick. Nevada has immense beds of this valuable mineral, which, by means of proper facilities. can furnish an ample supply of saltpeter to the country. Recently a large deposit of nitrate of soda was discovered in San Bernardino County, California, and it has also been found in the extreme southern por-

22

tion of New Mexico. The *carbonate* of soda abounds within the Great Basin, in extensive alkali flats, but it is always mixed with salt, borax, lime, magnesia, and other minerals. (*Min. Res. of U. S., 1883, pp. 597–602.*)

SULPHUR.

Sulphur, a simple mineral substance commonly called brimstone, is of a yellow color, easily broken, and burns quickly. The numerous mineral springs in different portions of the Union that have sulphur in their waters in combination with other ingredients give evidence of its general distribution. It is found in nature, as we have seen in treating of metals, everywhere in combination with their respective ores, so that its presence may be deemed almost universal. The States of New York and Virginia have each a number of sulphur springs, their waters having varied characteristics. We have also beds of pure sulphur, as in Cayuga County, New York; and on the Virginia side of the Potomac, some twenty miles from Washington; and on an island belonging to Ohio in Lake Erie; in Kansas, and in Florida. These deposits are all too small to have sufficient commercial importance to induce their being mined. There is, however, a remarkable bed of native sulphur in Southwestern Louisiana—some dozen miles from Lake Charles—that would seem to be a puzzle to geologists. This deposit lies 425 feet below the surface, and is about 100 feet in thickness, the sulphur being quite pure. Underneath this is also a bed of gypsum (sulphate of lime) about 150 feet thick. Texas has likewise beds of sulphur within her domain. In the Territories, sulphur occurs in Southwestern Wyoming, in Southeastern Idaho, and in Northeastern Utah, in the Uintah Mountains; these three beds appear to belong to the same general formation, though in Utah are a number of other localities where sulphur is found. New Mexico has also a number of beds of sulphur.

Pacific Slope Sulphur.—Beds of sulphur are found in a number of places in Southern California, along the west side of the Coast Range, and also in the southeastern portion of the State, in the vicinity of the mud-volcanoes in San Diego County. Some of these deposits are quite large, though only one, that at Clear Lake, has been utilized, but only partially.

Nevada has deposits of sulphur, but their percentage in richness is small. In that State it is often found in the openings of extinct hot springs amid a sage-brush desert. In Churchill County is a bed covering thirty acres, and some of the sulphur is quite pure, though somewhat mixed with earth. In the northwestern portion of the State is an important bed, which is evidently of volcanic origin, as it is in connection with *tufa*, a "volcanic sandstone," and the sulphur is imbedded in cavities, some five or six feet wide, with layers of crystals. "The sulphur has been derived from a deep-seated source, and deposited from the condition of vapor among the cooler and higher rocks where it is now found. Alaska has heavy deposits of sulphur near some volcanic cones along the coast, and in the groups of the Aleutian Isles.

The reason why our sulphur-mines have not been more utilized is because of the immense amounts that have come from abroad, especially from Sicily, where it is in abundance and where labor is exceedingly cheap, as well as the freightage.

Sulphuric Acid.—This acid is composed of one part sulphur and three parts oxygen; the common name is vitriol, or oil of vitriol. It is not strictly a natural resource, but it is easily obtained from the crude or natural sulphur, or from pyrites, or sulphurets—that is, when it is united in the mine with another substance as a metal. About one half of all this acid made in the United States is used in manu- ˙ facturing fertilizers, and, owing to the abundance of sulphur in numerous places, and combined in some form in nature, the acid is easily and cheaply obtained. Of one form it is said: " There is scarcely a State in the Union in which pyrites does not occur to a considerable extent; and at the present time all sulphuric-acid makers can be cheaply supplied from mines already opened," and there are hundreds of deposits to be drawn upon in the future. (*Min. Res. of U. S., 1883–1884, p. 880.*)

XXXII

MINERAL OR MEDICINAL SPRINGS.

THESE springs are found in several States of the Union, but by no means in all. The eastern portion, New York, and West Virginia, have the largest number; while, in connection with some of these, are pleasant resorts, especially for summer visitors, who are not invalids, but are seeking recreation.

In New England are a number of mineral or medicinal springs; the waters of many of these are deemed effective in remedying numerous ailments, though their flow is comparatively small. In Maine, at South Poland, are springs, the waters of which have a reputation as a remedy for kidney-diseases, dyspepsia, and kindred ailments. Near Milford, New Hampshire, are five springs, the waters of no two being alike. From one flows a water nearly allied in character to the famed "Apollinaris." The Sheldon or Missisquoi Springs, about twenty in number, are in Vermont, some ten miles northeast of St. Albans. These are the most important springs in the State. The waters possess different properties, but their principal virtue is claimed to be efficient in relieving cancerous diseases.

New York Springs: Saratoga.—The mineral springs in the State of New York predominate either in sulphur or soda. The most popular of these are at Saratoga, which is the most celebrated summer resort in the interior of the Union; but in the season there are present comparatively very few invalids. The region round about is not specially inviting as to scenery; the location may be termed a moderately elevated plateau, but the vicinity is re- markable for the purity of its air; the soil being dry and sandy, no malaria is generated. The number of springs and wells is nearly thirty, and they extend in a line northeast and southwest for a number of miles. The waters of no two are precisely alike, except that they are all more or less impregnated with soda, and are strongly charged with carbonic-acid gas. These waters are very diversified in their characteristics; one or two are chalybeate or iron; some have as ingredients magnesia, sulphur, or iodine. Their properties being in the main both tonic and cathar- tic, they stimulate the secretions and relieve the sys- tem, as in cases of dyspepsia or engorgement of the liver and kindred maladies. The waters of these springs flow in great abundance; being pressed up from the depths below by the force of gas, they come to the surface clear and sparkling. Though they have medicinal ingredients, they are nearly all pala- table and refreshing, and are drunk for their own sake, in great quantities, by the healthy visitors.

This State has in different localities a number of

other mineral springs; in these sulphur predominates more than any other ingredient. There are four springs at the village of Sharon, Schoharie County. The waters are diversified in character, there being one chalybeate, one magnesia, and two— one white, and the other blue—sulphur. These sulphur-springs are said to resemble those of Virginia; they are used for rheumatic ailments. In Otsego County are the **Richfield Springs**, at a village of the same name. Then quite a number of these are of varied size, but, upon the whole, the flow of water is not large; the predominant ingredient is sulphur.

Clifton Springs, in Ontario County, are quite famous for their healing properties in bilious, cutaneous, and rheumatic ailments. The important ingredient is sulphur, whose deposit is white; the sulphurous aroma so pervades the vicinity, that it can be detected at quite a distance. The sulphur combines also with other substances, such as magnesia, soda, lime, etc. The waters are similar to those of Greenbrier, West Virginia.

Avon **Springs** are in the valley of the Genesee River, south of Rochester. The waters of these springs have been characterized as "saline-sulphurous." They are used to relieve rheumatism, to aid digestion, and cure diseases of the skin. Lebanon **Springs.**—Their water, which flows from a cavity ten feet in diameter, is uniform in its temperature of about 73° at times, and flows in a strong current.

It is claimed to be remedial for nervous debility, liver-complaint, rheumatism, and cutaneous diseases.

Pennsylvania Springs.—This State can not boast of many medicinal springs, nor of their great value or flow of water. **Bedford Springs,** in Bedford County, have long been deemed the most important in the State. There are six in number; the prevailing ingredient in the water has been characterized as "saline-chalybeate," and also as containing sulphate of magnesia, carbonates of iron and of lime, and carbonic acid. The waters are tonic as well as laxative in their effects, and are remedial for dyspepsia and diabetes. The location is in a mountain-glen, and the altitude is such as to secure a delightful summer climate, and pure air that appears to invigorate the visitors as much as, if not more than, the waters themselves. The **Minnequa Springs** are located north of Williamsport, in Bradford County. Their waters are strongly impregnated with the oxide of iron, and are a tonic in their effect and a remedy for rheumatism and cutaneous diseases.

Cresson Springs, in Cambria County, bubble up curiously near the top of the Alleghanies, in the neighborhood of the village of Cresson, a summer resort located where the Pennsylvania Central Railroad crosses the mountains. There are in all seven springs, though they are not famed for possessing any very positive medicinal properties. The waters are impregnated to a certain extent with iron, and in one alum is found.

The Gettysburg Springs were visited, by invalids afflicted with diseases of the kidneys or bladder, long before the battle fought there had given the town a world-wide fame.

Fayette Springs.—These are in Fayette County, in an elevated valley encircled by the upper ridges of the Laurel Mountain. They are amid romantic scenery, where cool and pure mountain air may be enjoyed. The waters are not known as being specifically medicinal in their properties. (*Mineral Springs, p. 249.*)

Virginia Springs.—The White Sulphur, the most important of these springs, has for more than a century been a favorite watering-place. They are in Greenbrier County, West Virginia, near the western base of the Alleghany range. Taking the White Sulphur as a center, distant from them thirty-eight miles toward the north are the Hot Springs; the Sweet Springs are seventeen miles toward the east; on the south, twenty-four miles distant, are the Salt and Red Springs; and toward the west, twenty-six miles, is the Blue Sulphur. This constitutes what is often designated " The Virginia Springs regions "— extending north and south sixty-two miles and east and west forty-three. These springs are all amid mountains that afford beautiful scenery, while the elevation is such as to insure the exhilarating influence of pure air. We can not go into details, but there are in this region forty or more springs great and small. The waters as a general rule are charged with carbonic acid, and nearly all have sulphur as an ingredient and some also have alum. The waters are usually tonic and claimed to be remedial for neuralgia and rheumatism.

Blue Lick Springs. — These are in Nicholas County, Kentucky, and are in the valley of Licking River. The waters belong to the group known as saline-sulphur, and are deemed very fine of their class. They are efficient in relieving engorgement of the liver, gastric catarrh, gall-stones, and chronic diseases of the skin. These waters became known to the early settlers in consequence of their observing that the deer and the buffalo frequented the springs in order to lick the salt—hence the name ; the deposit is of sulphur, and beautifully blue in color. In the sandstone rocks, on the hill-sides bordering the springs, can be seen pathways, two or three feet deep, that were in the course of ages thus worn by these animals passing and repassing in single file, in their visits to the springs.

Wisconsin Springs.—This State has two localities in which are mineral springs ; both these are in the eastern portion of the State, and near Lake Michigan. The Sheboygan Springs have waters that resemble the Kissingen in Germany. The flow averages 225 gallons per minute from an artesian well that is 1,475 feet deep. The water is celebrated as a remedy for malarial fever, and also for kidney and liver affections. Directly south of these springs are the five springs near Waukesha. The most noted of these are the Bethesda and the Clysmie; their waters are effective in liver or kidney complaints, Bright's disease, etc., and also disorders of the stomach. Large numbers of patients visit these springs. (*Mineral Waters, p. 44.*)

The **Dakota Hot Springs** are situated in the southern portion of the Black Hills. The waters have an average temperature of 95° Fahr.; their medicinal properties relieve rheumatism and diseases of the blood and skin, and as such are found to be very effective.

Arkansas Hot Springs.—A little west of south of the springs in Wisconsin, and at a distance of more than 600 miles, are the Hot Springs of Arkansas. These are purely medicinal, and have remarkable healing properties. The temperature of the waters as they issue from the earth range from 95° to 150° Fahr., while the fifty-seven springs pour out this hot water at the rate of 350 gallons each minute. In the applications the patients use baths, and the results are very marked in relieving if not in removing scrofulous and rheumatic ailments. These springs, at an elevation above tide-water of 1,500 feet, pour their waters from the side of the mountain into a deep and narrow valley, almost a gorge. Patients congregate here from every portion of the Union.

Pagosa Springs.—New Mexico has a large number of medicinal springs in different parts of the Territory. The Hot Springs, in the vicinity of Las

Vegas, San Miguel County, are claimed to equal those of Arkansas in their healing properties. The Pagosa Hot Springs are on the borders of Colorado; in Rio Arriba County are the famous " Big medicine " of the Ute Indians. " The largest of these springs is at least *forty feet* in diameter, and hot enough to cook an egg in a few minutes. Carbonic gas and steam bubble up in great quantities from the bottom and keep the surface always in a state of agitation." The temperature of the water is 140° Fahr.; their altitude is about 7,000 feet. There are also a number of other springs in the immediate vicinity, besides a number in different parts of the Territory, as Ojo Caliente and Santa Fé.

Rocky Mountain Mineral Springs.—The numerous springs in the elevated valleys and parks of this entire mountainous region are claimed to compare favorably, in properties that are curative in kidney complaints, with those of the Hot Springs of Arkansas, or those of the waters of Waukesha, Wisconsin. (*Dr. Charles Dennison, p. 29.*)

The State of Colorado is rich in mineral springs, and has one locality where medicinal waters have afforded relief to great numbers of invalids. These are the **Manitou Springs.** They are amid the foothills of Pike's Peak, and at an elevation of 6,370 feet. They are six in number, and about eighty miles south of Denver, and five up the mountain from the village of **Colorado Springs,** on the railway. The Manitou waters, that is, of the several springs, are used in cases of "old kidney and liver troubles," and also in those of general debility. These springs are characterized as the "Saratoga of the West," as it is a popular resort for others besides invalids, the tonic properties of the atmosphere being an attraetion, and the waters, like those of Saratoga, can be drunk for their own sake.

The medicinal virtues of the waters of these springs were known for ages to the Indians in the vicinity, and they brought thither their sick that they might drink the waters and bathe in them. As they believed the relief which their sick received was a gift direct from the Great Spirit, they called the springs *Manitou.* The mineral springs in the Rocky Mountain region, though they may be somewhat different in their characteristics, amount to several scores in number. Among these in West Middle Colorado, in the valley of the Grand River, are Glenwood Springs, the waters of which partake of the general properties of those in that region. These springs are of recent discovery; but as access to them has become easy because of the junction of railways, the number of visitors has much increased.

California Springs.—This State claims to have hot springs the waters of which are remedial for rheumatism, gout, sciatica, paralysis, and cutaneous and blood diseases, while those of her cold springs are claimed to be beneficial in malarial, intestinal, and biliary ailments. The Napa Soda Springs contain magnesia, lime, iron, muriate of soda, with free car-

bonic-acid gas. The elements found in this water
are very similar to those in that of the famous Carls-
bad Spring in Bohemia. There are also many other
mineral springs in the State, of which about twenty
are in the vicinity of Calistoga. The waters of
these are hot, and are impregnated with iron, sulphur,
and magnesia. Within a radius of twenty-five miles
of the latter are a number of other springs which
partake of the general ingredients just mentioned.
Among these may be classed the famed Geyser
Springs in Sonoma County. Here are hot springs
and cold, quiet and boiling, sometimes within a few
feet of each other. " Some of the waters are clear
and transparent; others white, yellow, or red with
ochre, while others are of an inky blackness"; they
also differ as much in taste and smell : some are impregnated with alum and salt, and others are sulphurous and fetid in odor. They afford almost a
certain cure for gout, rheumatism, and skin-diseases.

The Gilroy Hot Springs, eighty miles south of
San Francisco, are on the west slope of the Coast
Range. They vary in temperature from 109° to 115°
Fahr., and have as ingredients sulphur, alum, iron,
iodine, and magnesia, with a trace of arsenic. There
are in California more than forty localities where
mineral springs exist, and each of these places has a
number.

In many sections throughout the Union are found
isolated medicinal springs, that have a local and uncertain reputation, because of cures which they are

said to have performed. As a general rule the flow of water from these springs is so very limited that, even if their remedial properties were valuable, they are, in consequence of this deficiency, of little practical utility.

XXXIII.

HEALTH-RESORTS.

IN connection with mineral or medicinal springs are sometimes other attractions that make their vicinity a resort for those who seek recreation and amusement, rather than rest from physical and mental toil, or both combined. We have in the United States a number of localities that are purely health-resorts, and which derive their character, as such, either from climatic influences or from elevated loca-. tions, that possess an atmosphere free from impurities, and sometimes from the union of both these elements. Since the human system is liable to become weakened by the inroads of certain diseases that are slow in their progress, such as nervous debility in its numerous forms, and pulmonary complaints in their various grades, it would seem that remedies that are correspondingly slow in their action are very often effective. Thus, changes to more genial or invigorating climates are frequently found beneficial, and, though the perfect restoration to health may not always be attained by the eradication of the disease itself, yet these mitigating influences may smooth the pathway of the invalid, perhaps, during many years. In this manner health-resorts are fraught with blessings to multitudes of people living in widely separate sections of the country, who can thus change their residence, in order to secure health. Other advantages of health-resorts can be enumerated, inasmuch as in our Government the people themselves are deeply interested, since, being voters, they have a certain responsibility; and, while there is danger that, in a territory so extensive, diversified interests may clash, yet an antidote to such evil is present in the frequent intercourse among the people of the different sections, which has the effect of removing prejudices, and of inducing a sentiment of nationality and sympathy as between the members of the same Nation.

Health-Resorts east of the Alleghanies.—Many localities in the White Mountains, because of their altitude, have the reputation of affording relief to those afflicted with asthma and hay- or rose-fever. Similar places are said to be found in the Catskills and in the Alleghanies; notably at Altoona and Cresson in Pennsylvania.

On the southern Atlantic slope are three prominent health-resorts: *first*, in certain localities in Florida; and, *second*, in the highlands of Western South Carolina; and the *third*, in Southwestern North Carolina. The climate of Florida is unusually affected by its surroundings of water—the most important in influence being the Gulf Stream.

23

This flows along its eastern coast, and imparts to the land atmosphere a portion of its own warmth of air, so charged with vapor that, though the temperature may be high, there is no indication of parching dryness. Even in winter the atmosphere is balmy and pleasantly invigorating. The winds coming from off the ocean are the northern surplusage of the trade-winds, and are usually from north of east, while those at certain seasons coming from the southwest are moderated by the waters of the Gulf of Mexico. The climate is, therefore, remarkably uniform, there being no sudden nor great changes.

Great numbers from the more northern portions of the Union resort to Florida in search of health; among these are included those also who are not absolutely invalids, but their health and constitutions have become so much impaired that they seek a milder climate than is afforded by chilling snows and bleak winds, and in Florida they find the needed tonic in an air deliciously fresh and balmy, and pleasantly fragrant with the perfume of flowers. Thousands of these invalids seek relief, especially from pulmonary complaints, and perhaps as many more for relaxation for a time from the wear of mental toil or from the harassing cares of business. Nor must it be overlooked that this pleasant region can be reached within a day or two from the snow and ice and bleak skies of the North.

Covington.—In connection with Florida may be

noted a district in Louisiana which partakes of many characteristics of an extreme Southern health-resort. Some physicians of New Orleans claim that peculiarity for a district north of Lake Pontchartrain, of which the village of Covington may be deemed the center. The region lies comparatively high above the coast-level, and in the midst of pine-woods; the soil is dry, and the vicinity is sheltered from "northers," which otherwise in that section of the country sometimes burst upon the land. The air is soft and free from the usual dampness from off the Gulf, and in the winter months the region is very healthful. It is claimed by some to be almost a specific in "its soft and piny atmosphere" for pulmonary diseases or bronchial troubles, healing diseased lungs, and restoring worn-out humanity.

Aiken.—In Western South Carolina the village of Aiken may be taken as near the center of the health-resort area of the State, which, according to an authority (*Handbook of South Carolina, p. 122*), is 2,000 square miles. The general temperature is very nearly the same as that of Santa Barbara, California. The atmosphere of the region is remarkably dry, but bracing. The elevation—700 feet above the ocean—and the subsoil being sandy, render the earth dry, while the waters of the springs and wells are very excellent because of their pureness. The prevailing winds are from the south and southwest, and the climate during the winter may be termed balmy, there being, as a rule, scarcely any snow, and that melting almost immediately after reaching the ground. "These sanitary conditions are present with the terebinthinate (qualities of turpentine) and the healing odors of a great pine-forest." This region is sought by numbers of those afflicted with sensitive lungs, and who wish to avoid the harsh winds of the winter season in the North.

Asheville.—In a west of north direction from Aiken, about 140 miles, is another health-resort area, but whose characteristics are different, its elevation being more than three times that of Aiken. The central position of this area is in the vicinity of the pleasant village of Asheville, in the county of Buncombe, in Southwestern North Carolina. This village is 2,250 feet above tide-water, and is situated between the Blue Ridge and the Alleghanies, "in the lovely valley of the French Broad," a tributary of the Tennessee. "It is surrounded by an amphitheatre of hills, commanding one of the finest mountain views in the Union." This district having been so much frequented as a health-resort, it has been found expedient to establish there "a sanitarium for consumptives, the sunny hills of this locality offering ample facilities for out-door life." This entire region has a remarkably equable climate, mild and dry, and specially adapted to mitigate the ravages of pulmonary complaints, and, in consequence, here are found such patients both summer and winter.

Walden Ridge.—West of Asheville, a hundred miles or more in East Tennessee, amid the Cumberland Mountains, are table-lands some 2,000 feet above the ocean, where the inhabitants are virtually exempt from pulmonary diseases, "consumption being almost unknown among the natives." This plateau is known as Walden Ridge. It is quite elevated, the air is remarkably pure, and, in the opinion of the physicians who have studied the subject, this entire region has the elements that constitute a valuable health-resort. "Southern Georgia supplies in winter a sanitarium for pulmonary diseases, and Northern Georgia in summer for malarial diseases and fever; indeed, for lung-diseases also." (*Commonwealth of Georgia, p. 12.*)

A **Dry and Cold Climate.**—It often appears, in pulmonary complaints, that some victims find the heat debilitating, in a southern and warm climate, though the air may be fresh and pure, and they require a cooler atmosphere, but one equally pure. This requirement is found, perhaps, more perfectly in the State of Minnesota and the adjacent region, than elsewhere in the Union. The entire elevated region—whose average altitude is about 1,200 feet— lying west of Wisconsin to the Pacific slope, when fully tested, may be found favorable as health-resorts for pulmonary diseases. The severe cold and bright skies in winter, with a dry and fresh atmosphere, render the climate of Minnesota very exhilarating, and an almost certain antidote to lung-diseases in their incipient stages. This is a natural result of strengthening the system, because the dry, bracing air induces an appetite for plain, substantial food, which appears to be more easily assimilated in a cold than in a warm climate, and thereby the system of the invalid is invigorated and vivified with new life. That this climate is bracing to the human constitution, and never or seldom debilitating, is the testimony of those who have practically tested its properties. "Very many of those now residing in the State were induced to come hither because of bronchial or pulmonary ailments, and in all cases relief was experienced, and most of them have fully ·recovered." (*Minnesota Illustrated, p. 14.*) The average annual death-rate in the United States is one in 74; in Minnesota it is one in 112. The latter includes invalids who were from outside the State, and, suffering with chronic diseases, came there to obtain relief "when their vitality was so far exhausted as to render recovery scarcely possible."

A Dry Climate; Altitude and Sunshine.—These three elements, when combined with a mild and equable temperature and pure air, make a paradise for invalids, especially those who are afflicted with lung-diseases. The latter are the most difficult of all chronic complaints to alleviate, much less to cure, because the organs wherein the disease is seated can not possibly have rest or time to recuperate, but must be kept in motion constantly. The three conditions mentioned above are found to a great degree in Colorado, especially in the southern portion and on the plains in the vicinity of Denver. The climate there is comparatively dry, as the clouds from off both the Atlantic and the Pacific are nearly exhausted before they reach that region; cold days are few in number, and snow falls but lightly and remains only for a short time, and the days of continuous sunshine and cloudless skies are in the great majority. "According to the records of the Signal-Service office at Denver, there were only *seventeen* days from January 1, 1873, to September, 1878—five years and nine months—in which the sun was invisible throughout the whole day." (*Dr. Dennison, p. 72.*) Says another authority: "No one need be afraid of the sunlight of Colorado. It has all the good effects of sunlight in other countries, with none of its enervating influence, so common elsewhere." Let the consumptive bask in it and enjoy all he can "the beneficial effects of the chemical action of sunlight on the blood. . . . While desirable coolness increases the oxygen-containing capacity of the atmosphere, altitude has, also, a counter-influence, and necessitates an *active out-door life* to insure the best results." Though the temperature is not so mild as that of Florida or Southern California, and though the climate differs in being influenced by a greater altitude, yet in the main it is genial to consumptive invalids, and experience shows that the disease in its earlier stages may be cured; but, if otherwise, the patient can be greatly relieved. The altitude of Denver is

5,200 feet, and any desirable greater altitude can be reached within 100 miles.

The Territory of Utah lays claim in many respects to be a health-resort for consumptives. Says a United States surgeon: " In an experience of three years and a half in Utah, I have not seen a case of consumption that originated in the Territory." In respect to freedom from lung-diseases of the native residents of New Mexico and Arizona, portions of which in the near future will no doubt become health-resorts, says Dr. Irwin, of the U. S. Army, as quoted by Dr. Dennison: " During a seven years' residence in New Mexico and Arizona, I never saw or heard of a case of tuberculous disease among the native inhabitants of those Territories."

Asthma.—In this connection, we may say that all the elevated region including Colorado, Utah, New Mexico, and Arizona is a vast house of refuge for those afflicted with that wearisome and painful affection—the asthma. Thither come multitudes to be relieved. Recently a record of these invalids was made, and it was found that they had come from *twenty-three* States.

A Balmy Climate near the Pacific.—The southern portion of California may well be described as a universal health-resort. In a curve of the southwest foot-hills of the Coast Range is located the old town of Santa Barbara. This is known as one of the most desirable and most frequented of any of the resorts in that region. Many of its climatic advantages are due to its peculiar situation, it being nearly encircled on the north and northwest by the foot-hills and mountains from three to four thousand feet high, and on the south and southwest by the Pacific, upon which it looks out from its sheltered nook. The town is distributed over about three square miles, so that there is plenty of breathing-room for the citizens. The population is about 6,000, and it is estimated that one half have come thither to find shelter from disease, and to seek health. Great numbers of these have come from the New England States, and generally from the northern portions of the Union, and principally on account of consumptive ailments, of which they wish to free themselves. This whole region is deservedly popular because of its health-preserving properties, and multitudes go there who are not blessed by Nature with strong constitutions, and must seek a less harsh climate—some to remain permanently, and others only until their restored health will authorize them to return to their original homes.

Santa Barbara is a charming locality : orange-trees are found everywhere ; roses abound both winter and summer. " Verbena-beds are cut down like grass thrice a year, but spring up again stronger than ever ; the heliotrope climbs twenty feet high upon a support." In **Los** Angeles great numbers of the inhabitants are temporary residents, some for pleasure, but more in pursuit of health, especially those affected with pulmonary complaints. **San**

Diego, about 130 miles southeast of Los Angeles, is claimed to have "the mildest and sunniest winter climate on the coast." This locality is not at present so accessible, but it is equally as charming and health-preserving. The climates of Santa Barbara, Los Angeles, and San Diego, because of the great humidity of the atmosphere, have been compared by writers to the warm and moist climates of the basin of the Mediterranean. This balmy and health-giving influence extends down the coast to and beyond the Mexican line.

Yellowstone Park. — This park, which is set apart by Congress for the use of the people of the Union, gives indications of becoming a valuable health-resort as well as one for recreation. Prof. Edward Frankland, of England, an eminent physician and scientist, in "The Popular Science Monthly," July, 1885, says: "The great importance of a winter sanitarium for patients suffering from or threatened with consumption and other allied diseases has long been recognized and acted upon in Europe." He then compares the favorite health-resorts of Europe, in the secluded valley of Davos, in the Engadine, in Switzerland, with that of the Yellowstone Park, saying that the latter "rivals if it does not surpass Davos in the excellence of its winter climatic conditions." The elevation of Davos is 5,400 feet above the sea, that of Yellowstone Park is between 7,000 and 8,000. He continues: "Invalids could remain in the dry atmosphere of the latter till the first of May, while the moisture at Davos from melting snows compels them to leave early in March. . . . It is to be expected, from this greater elevation, that a clearer sky and a larger proportion of sunny days would be experienced in the Yellowstone Park, while the wholesomeness of the air would be still more marked, owing to its comparatively greater freedom from *zymotic matter.*"

Dr. Frankland, after investigation, thus sums up in respect to the Park : " Dedicated during the winter months to the purposes I have advocated, it would constitute a winter sanitarium unequaled in the world, restoring to health and vigor, not only persons suffering from incipient chest-disease, but also those of the overworked populations of the States and of Canada."

XXXIV.

RAINFALL—OCEAN-CURRENTS.

THE United States possess a natural resource in a soil, taken as a whole, uniformly fertile ; but what would be its worth, were it not stimulated by the bright sunshine and the abundant rainfall, to put forth its energies in growing and maturing crops? The intelligent reader will be interested in tracing causes of this copious rainfall, while he recognizes the sunshine as a gift direct from the same beneficent Creator.

As we ascertain causes and trace their influence, how clearly becomes the evidence of the Creator's design, as manifested in the mode by which He has made provision for the requisite moisture to vivify the productions of the earth, and also provided for warming its colder regions and for cooling those that would be otherwise intensely hot—and all by using the currents of oceans, whose movements He has established !

In considering the ocean-currents we will notice their effect upon the climate and rainfall of our own country, as it will be seen that North America, and

The Means to an End.

especially the United States, owing to their central
position in that continent, derives as much if not
more benefit from the two currents—the Japanese in
the Pacific, and the north equatorial in the Atlantic—than both Asia and Europe combined. If the
lands in the southern hemisphere were as extensive
as those in the northern, there could exist no inexhaustible reservoir of water like the Antarctic Ocean.
Eternal desolation and dryness, induced by cold,
would reign over the greater portion of both the
northern and southern hemispheres, while within the
tropics, with scarcely an alleviation, a dryness induced by intense heat would hold sway. To man, as
at present constituted, such climates would be intol-
crable.

The Means to an End.—The arrangements designed to secure the desired ends seem to be perfect. The extreme of one continent (South America) penetrates this great Antarctic reservoir like
a wedge, on either side of which currents of water,
in consequence of the centrifugal force generated by
the rotary motion of the earth, rush toward the
equator, under similar conditions and force, as did
originally the surface portions of the mass of the
earth itself, when, in its fluid state, it was sct revolving on its axis. This fluid mass of earthy matter
continued thus to flow from either pole until along
the entire length of the equator, as a middle line
around the globe, was accumulated a protuberance
—its greatest depth being thirteen and a quarter

miles on that line. This fluid mass finally hardened into the present permanent form. There is no evidence that the rate of the earth's revolution on its axis has either been increased or diminished since it received its first impulse.

Currents of Water and of Air.—The main current, coming up on the east of South America, divides itself on the Cape of Good Hope; thence a portion goes into the Indian Ocean, and, because of its coldness and passing out slowly, gives the latter a lower temperature, thus mitigating the heat of the surrounding lands. The other portion of the current, in which we Americans are more interested, sweeps up along the west side of Africa, meanwhile moderating by its coolness the temperature of the latter's southern part, which otherwise would be "fiercely heated by the rays of a tropical sun." On approaching the equator it comes still more under the influence of the rotary motion of the earth, and its course is gradually deflected toward the northwest and finally directly west, and thus continues across the Atlantic. The earth, in its revolution eastward on its axis, literally turns under the water, which does not fully get its motion, but is left, and practically moves toward the west. The trade-winds are attributed to the same cause. The colder and denser air of the extremes, north and south, presses toward the warm tropics, where, becoming rarefied and lighter because of the heat, it floats upward to the higher regions of the atmosphere, thus making a partial vacuum, to fill which the cold and comparatively heavy air from the poles rushes in, to be in turn rarefied and set afloat; this process goes on unceasingly. The weight of the air is lighter in the tropics than in the temperate zones; for that reason the height of the barometer on the equator at the level of the ocean is less than anywhere else. Experiment shows that at 10° latitude, either side of that line, the barometer is nearly 4° higher, owing to the greater density of the atmosphere. The natural course of these winds coming from the north and south poles toward the equator would be on meridian or direct lines, were it not for the earth in its rotary motion running against them, and, somewhat like a railway-car in motion creates a breeze, while also changing their course, so that north of the equator they first blow from the northeast, and south of the same line from the southeast, but, soon blending together on one common line, they press directly west across and beyond the Atlantic. The air being much lighter than water, these trade-winds acquire a correspondingly greater velocity than the currents of the latter. The winds return only partially into themselves, they having the circuit of the earth for their limit; but the ocean-currents do, because their boundaries are the shores of the adjoining continents, which obstruct their flow.

Origin of the Gulf Stream.—Let us briefly notice the principal current of the Atlantic, which furnishes so much heat and moisture to Western Europe, while at the same time it is indirectly equally essential in conferring a copious rainfall upon the interior of our own country. At a point on or near the equator a little west of the Gulf of Guinea, this current—the Atlantic equatorial—takes a due west course, as there the impetus received from the motion of the earth has its full influence and impels it in that direction. (*Appletons' Physical Geography, p. 51.*) The first ob-struction in its path is Cape St. Roque, South America, upon which it divides; a portion passing to the southwest along the coast, and which is known as the Brazil Current, but the much greater portion is deflected toward the northwest. The latter sweeps along the northeast and the north coast of South America, carrying with it evidently a large portion of the waters of the Amazon, the Orinoco, and the Magdalena—all warmed by a tropical climate. It continues on westward through the Caribbean Sea; a large portion of the current passing northward through the Yucatan Passage to a point in the Gulf of Mexico opposite the straits between Florida and Cuba, where it turns eastward through the latter straits. That point is 500 miles west of the straits, and a little greater distance southeast from the mouth of the Mississippi.

Here also this current absorbs a large portion of the warm waters of the Rio Grande, of the Mississippi, and of the Mobile and other rivers. The waters of these rivers flowing into the Gulf of Mexico must have an outlet somewhere; there is only one, however, and that is made available by means of their blending with this ocean-current that comes up through the Yucatan Passage, while they are somewhat diminished by evaporation. Finally, this current rushes out into the Atlantic between Florida and Cuba, at the rate of nearly five miles an hour, and in a volume more than forty miles wide and nearly 3,000 feet deep. (*Appletons' Physical Geography, p. 60.*) Now, under the name of the Gulf Stream—after being joined east of the straits by a similar warm ocean-current flowing west along the north coast of Cuba—this mass of waters starts toward the northeast on an errand of blessings for the western shores of Europe. At first the onward movement is so rapid that the warm water, as it naturally would, has apparently not time to rise sufficiently high to overflow, but forces its way through the surrounding waters. In advancing the stream deflects gradually from the coast, its rate of speed diminishes, and its waters begin to expand over the surface of the ocean, while its depth grows less and less. Opposite Cape Hatteras the stream swerves nearly one hundred miles from the coast and is seventy-five miles wide and much diminished in depth, though its surface temperature is scarcely changed. This immense volume of water, estimated to equal a thousand Mississippis at full flow, has a temperature on the surface, when leaving the Gulf, ranging from 80° to 82° Fahr. The aggregate heat thus obtained is in consequence of its passing under

24

a tropical sun more than 4,000 miles, including the curves in the Caribbean Sea and through the Yucatan Passage, together with the accumulation of the warm waters from the rivers mentioned, so that the stream through the greater part of its course is on an average from 10° to 15° warmer than the surrounding waters. The water of the Gulf Stream has been described "as clear as crystal and intensely blue."

Effect on Climate.—The nearness of the stream to the coast of the Florida Peninsula has a modifying influence in producing in that region a climate unusually equable; but as it recedes from the coast the northeastern portion of the Atlantic slope derives from it little if any benefit, since the prevailing winds in that region come from the west or northwest, and drive toward the east the warm vapors rising from off the Gulf Stream, while in addition there flows from the northeast a cold current between that stream and the coast. " In the North Atlantic the west winds prevail to such a degree, that the average passage of fast-sailing vessels from America to Europe can be made about two thirds shorter than on the same route from Europe to America." (*Prof Guyot, Earth and Man, p. 98.*)

The Compensation.—But for this lack of warmth and moisture derived from the Gulf Stream on the middle and northern part of the Atlantic slope, ample compensation is given to another and much larger portion of the Union—the basin or valley of the Mississippi. The latter section derives a bene-

fit in its rainfall, almost infinitely greater to the nation than if the influence were felt only on the territory east of the Alleghany range of mountains. In
addition, a large amount of the moisture that blesses
the foot-hills and intervening valleys lying along the
eastern side of the latter range up to the parallel of
42° is derived principally from the surplusage or
overflow of the vapor-loaded winds from the valley of
the Mississippi, and which move east over and
through the mountains. The great benefits thus
conferred upon the United States by the Atlantic
equatorial current are derived indirectly from the
vast evaporation of water from off its surface.
This current is about 3,000 miles wide north and
south, and moves toward the west at the rate of
thirty or thirty-five miles a day for more than 4,000
miles, and while thus under a broiling tropical sun
its waters are heated, on an average, from 80° to 82°
Fahr.

During this passage there is continually going on
a steaming process of evaporation by which it is estimated that " off this ocean-belt there is in the form
of vapor annually floated up into the higher air fifteen feet of water." (*Maury's Geography of the Sea,
p. 102.*) " The sun causes these invisible vapors to
rise, which, being lighter than the air itself, increas-
ingly tend to soar into the upper atmosphere, filling
it, and constituting within it another aqueous atmosphere." (*Prof Guyot, Earth and Man, p. 85.*)
These vapors in their ascending movement encount-

er the colder layers of the higher regions of the air, and thereby they are somewhat cooled, and under the form of clouds or fogs are borne along by the winds, and finally descend in rain to fertilize and cause the earth to bloom.

Sources of Rainfall in the Valley.—The trade-winds, saturated, as we have seen, by invisible vapors, rushing west across the Atlantic, meet their first obstruction in the plateau of Mexico, and especially in the Sierra Madre Mountains on its western border. The latter are on an average about 9,000 feet above the sea—nearly two miles—and stand across the path of these saturated winds at an angle that deflects them toward the north. Thus checked in their onward progress, the different strata of this air in motion are shoved up one upon another, because of the continuous pressure from the east. The upper currents surge high into the atmosphere, even so far that the influence of the rotary motion of the earth must be quite diminished, while the natural attraction toward the north pole to restore the equilibrium, causes this air to flow in that direction, as on an inclined plane, over the colder and therefore denser strata of the air beneath, till it reaches the surface of the earth. That point in summer is about 30° north latitude, but—by a wonderful provision—in the winter, when the surface air is colder and more dense, the earth is reached much farther north, where the warmth is especially needed. These winds, still holding vapor or water in solution, and flowing in this direction, are characterized as the "anti- or re-turn-trades" by meteorologists.

The Trade-Winds deflected northward. — It is evident that these trade-winds of the tropics do not pass over the Sierra Madre or Mexican Mountains, and thus continue their westerly course, as there is no indication of their presence on the west side of that range, nor on that portion of the ocean immediately adjoining the coast. On the contrary, the effect of the revolving motion of the earth on the atmosphere becomes perceptible only at about one hundred and fifty miles west of Mexico, for near that point the Pacific trades have their origin, similar in manner to that of the trades of the Atlantic ; the latter having their origin at about the same distance west of the coast of Africa. Says Prof. Orton (*The Andes and the Amazon, p. 118*), in citing an analogous case, that of the Andes : " So effective is that barrier that the trade-winds are not felt again on the Pacific till you are one hundred and fifty miles from the coast." It is clear that the north flank of these Atlantic trades thus obstructed, and meanwhile pressed from the east, must find vent somewhere ; they can not pass toward the south, because of the greater force of the main current, and as we have seen they do not pass over the Sierra Madre, it follows that their only outlet lies toward the north, and in that direction they must necessarily be driven. That such is the case is confirmed by the fact that the angle at which they strike the mountains would deflect them almost due

north. Mr. G. W. Featherstonhaugh, U. S. geologist and explorer, expressed the opinion, nearly half
a century ago, that these saturated trade-winds, in
being deflected north by the Sierra Madre Mountains, furnished the valley of the Mississippi with its
rainfall.

The Gulf a Source of Rainfall.—It has been virtually assumed that the evaporation off the waters of
the Gulf of Mexico alone supplies the rainfall of the
valley of the Mississippi. " By far the greater portion "—of the rainfall, meaning for all the country—
" comes from the Gulf and spreads over the central and
eastern part of the Mississippi Valley, and even much
of the Atlantic plain " or slope. Again: " The warm
southerly air-currents, loaded with moisture from the
Gulf, pass up the Mississippi Valley," etc. Here the
influence of the return-trades, heavily saturated with
vapor and driven north, as we have seen, from the
Sierra Madre, are ignored. Let full credit be given
the evaporation off the Gulf for its share in producing the rainfall in the valley; but that it could alone
supply that rainfall is a question open to several
objections. One of these is absolutely insuperable—
that is its inadequacy to produce sufficient moisture
for the purpose. To illustrate: The area of the Mississippi Valley is estimated at 1,244,000 square miles,
and the area of the Gulf as one fourth as much. The
annual average rainfall in the valley is forty-two
inches — three feet and a half; to produce this
amount would require an evaporation of *fourteen feet*

of water off the surface of the Gulf, presuming that it was all thus utilized. It is obvious that so great an evaporation is impossible. This estimate does not include the large surplusage that passes from the valley over and through the Alleghany Mountains, to eke out the deficiency of rain on their eastern slopes, as derived from the Atlantic Ocean, because the influence of the latter's vapor-loaded winds appears not to extend usually farther inland than about one hundred miles.

Taking the flow of the Mississippi as the standard, we may safely estimate that the amount of water annually poured by rivers into the Gulf would raise its surface—if there were neither evaporation nor outlet—one foot and a half. (*Appletons' Physical Geography, p. 130.*) If that amount of water were evaporated annually, there would still be *twelve and a half feet* to be accounted for. The surface of the Gulf, however, remains uniform in its height, and it has been discovered recently that the portion of the Atlantic equatorial current, that comes up through the Yucatan Passage, penetrates the Gulf only about 500 miles in passing onward to the Straits of Florida. There appears no evidence that, in thus passing through, this current contributes any water to that of the Gulf. The unusual annual rainfall—about sixty inches—on the north shore of the Gulf, and for one hundred miles or more inland, is to be accounted for, because, at the parallel of 30° north, the lower strata of the "return-trades" reach the earth, and there mingle with the vapor-bearing winds off the Gulf. It is evident that a large portion of the flow of the trade-winds in crossing over the Gulf passes by the mouths of the Mississippi. This is an inference from the fact that the sea-breezes off the Gulf penetrate unusually far into Texas and the adjoining regions of Mexico. To furnish the amount of water sufficient for an annual rainfall of such depth, and over a surface so extensive as that of the Great Valley, would certainly require an evaporation off a surface as great as that of the northern portion of the Atlantic equatorial current.

The Course of the Arctic Winds.—It is interesting to note the utilizing effect of the cold, dry north winds, when they come in contact with the warm vapor-loaded winds from the south. In the region, north of the valley and west of Hudson Bay, the direct flow of the winds from the Arctic Ocean toward the Gulf of Mexico is aided by two causes: one, the Rocky Mountains, trending southeast, which break the force of the west winds that would interfere with the southern direction of the former; 2. The average slower rate of the earth's rotary motion in that latitude—from 46° to 60°—which is not quite two fifths of that at the equator. The latter motion is not sufficiently rapid to have much effect in diverting the course of such winds toward the west, as are the trade-winds of the tropics, and therefore in this high latitude these winds for the most part flow virtually due south till they impinge upon

the eastern face of the Rocky Mountains, and as the
angle of incidence is equal to the angle of reflection,
these winds, though diverted easterly, are still attracted toward the equator and move in that direction.
Says Prof. Guyot (*Earth and Man, p. 100*): "The
polar winds, seeking the equator, strike obliquely
against the Rocky Mountains, and in running along
their eastern slopes are deflected to the southeast
and become the northwest winds of the valley of the
Mississippi." Are these the original cold waves reported so often by the Signal - Service Bureau?
Again: "These cool winds meet the surplusage of the
moist return trade-winds, and by their coolness con-
dense still more the latter's vapor which descends in
rain-storms, that are sometimes quite violent, but
furnish the water for the head-streams of the Missouri and its branches."

It may be added that the influence of these
winds extends much farther south within the valley,
because, farther to the east, for nearly 800 miles, they
are not thus impeded by the highlands or mountains,
but pass over the low plains of the Saskatchewan
and north of Lake Winnipeg, and, crossing the latter,
a portion find their way unimpeded up the valley of
the Red River, and easily pass over the comparatively low divides into the valleys of the Missouri
and the Mississippi. Meanwhile another portion,
still farther east, enters the upper end of the latter
valley, down which they pass. These cool winds
coming from the north are remarkably dry, but now

they come in contact, and mingle with the " return-trades," that have been saturated with moisture, as we have seen, from off the Atlantic equatorial current. The effect is that the vapor of these return-trades is condensed into mists and clouds, that eventually descend in copious rains.

Comparison of River-Valleys.—The basin or valley of the Mississippi — that is, the entire region drained by that river—is a most important factor in its bearings upon the material progress of the Nation. The fertility of its soil, its varied climate, its sunshine and rainfall, the variety of its productions, from the cereals and orchard-fruits of the northern portion to the cotton and sugar and semi-tropical fruits of the southern, all of untold value in supplying the wants of the people, make this valley strikingly superior to any other in the world. Its great area extending north and south, and being open at both ends, the flow of the winds is not impeded, while they greatly modify the climate and supply moisture. The height of land, or divide, near the parallel of 49° north latitude, between the mouth of the Mississippi and the Arctic Ocean, is only about 1,600 feet above the sea, and that not abrupt, the approach being so gradual on both sides that the winds pass over easily.

The only valley in the world that compares with that of the Mississippi, both in size and position, is that of the La Plata, South America. But what a contrast in respect to climate! The former is never affected by extensive and disastrous droughts, but is wonderfully uniform in its rainfall, while the latter has what is termed its rainy season, when Nature vigorously puts forth her energies and covers the vast plains with rank and abundant grasses, on which feed innumerable herds of cattle, horses, and sheep; afterward come the droughts, in which this grass withers beneath the broiling sun, and the plains once so green are covered inches deep with dust. "These terrible droughts on the pampas or plains in the valley of the La Plata cause the cattle as well as wild animals to perish by thousands and millions. . . . Before they starve the wretched creatures consume not only the grass, and every vestige of vegetation, but the very roots, which they paw out of the earth with their feet." These droughts are often succeeded by locusts that come up " like storms in vast and dense purple clouds in the distant sky." When wearied by flying they fall upon the ground, sometimes "ankle-deep," and devour what is left. (*Gallengass's History of South America, pp. 283, 284.*)

Comparison of Plains.—There is an equally striking contrast, which may be noted in this connection, between the plains of the La Plata and the high Western plains of the United States. The grass upon the former is so rank that its nutritive properties are thereby much diminished, while upon the latter the short and compact buffalo and bunch grasses are famed for their remarkable richness in these qualities. Upon the former the grasses wither and die under a broiling sun, but on the latter, while still standing, the sun cures the grasses and they become hay, but retaining their original nutritious properties. On the Western plains, instead of burning suns, come snows, yet the cattle and buffalo live and fatten on this cured wild hay ; they remove the snow with their hoofs, not to eat grass-roots, but food rich in nourishment.

Comparison of Continents.—It should be deemed a natural resource of immense value to the people of the United States that their territory lies between the two oceans—Atlantic and Pacific. The advantages of this position consist in two respects : one in relation to international commerce—of which the scope of this volume does not permit discussion; and the other pertaining to climate and rainfall, the latter two being peculiarly influenced by the great equatorial currents of both these oceans. The distance from ocean to ocean across the United States is in contrast with that of Europe and Asia combined; for, though the latter are designated on the maps separately, their territories are united. Extreme portions of Asia are so far distant from the sources of its rainfall, whose origin exists in the vapors that rise from off the surrounding oceans, that immense areas in the interior are absolutely barren or desert, inasmuch as the moisture which at first saturated the winds becomes exhausted before the latter reach these sterile districts.

On the parallel of 40° north latitude, the difference between the combined width of Europe and Asia, and that of the United States, is about 5,000 statute miles, and on that of 44° it is about 4,500. The width of the United States on the 32d parallel is 2,183 statute miles; on the 40th, 2,703; and on the 42d, 2,754; while on their southern border for 915 miles is the Gulf of Mexico, which is open to the northern flank of the trade-winds of the Atlantic, and as a sea-breeze these winds extend unusually far into the adjoining territory. On their northern border, also, lie the Great Lakes, having an area of nearly 100,000 square miles, more than half of all the fresh water on the globe. Owing to the course of the winds in that latitude, the much greater portion of the vapors rising off the surface of these lakes is carried within the United States.

Direction of Mountains.—The mountains of Asia, especially the Himalayas—the highest in the world —in their general direction lie east and west, thus impeding the natural flow toward the north of the warm winds saturated with vapor off the Indian Ocean, and which would modify the climate; equally for the same reason the cool winds from the north are cut off: hence the intense heat on the plains of India. The same, or nearly so, may be said concerning some of the ranges of mountains that run east and west on the north shore of the Mediterranean, as they deprive to a limited extent certain districts of country of the warm winds coming across from Africa, and which in their passage become saturated with moisture which they have taken off the surface of the sea.

The mountain-ranges of the United States, on the contrary, in the main run north and south, thus affording ample facility for the winds saturated with vapors off the Atlantic equatorial current to find their way, as we have just seen, in a northerly direction, over a vast extent of territory. Neither does the altitude of these mountains prevent entirely the passage over them of the surplusage of these vapor-loaded winds.

The Absence of Deserts.—The American people will notice that the territory of the United States contains no deserts, in the sense in which such immense barren wastes are found in the Old World. There are, it is true, certain small districts, when compared with the great mass of their domain, that are sterile and unproductive, such as the Staked Plain of Texas—though it abounds in fertile valleys and rich localities of large extent, which are well watered—and small portions of Arizona and New Mexico. The latter compensate the Nation by their mines of precious metals. Theory assumes that the return-trades pass to the east of these comparatively dry districts in Arizona and New Mexico, while the high mountains on the west prevent the vapor-loaded winds from off the Pacific reaching the dry and sterile portions of Nevada and Utah. There are also two sterile districts in Southern California; they lie on both sides of the San Bernardino Mountains, which trend northwest-southeast; the one on the northeast side is called the Mojave Desert, and the one on the southwest the Sandy or Colorado, as it extends from the river of that name. These districts combined are estimated at more than 10,000 square miles.

The Great American Desert.—At one time on the maps of the Union could be seen a wide and long stretch of territory, extending north and south, and lying east of the Rocky Mountains, on their slope toward the Mississippi, which was designated the "Great American Desert." Strictly speaking, there is no desert land within the valley of the Mississippi. In more recent times a United States army officer, a graduate at West Point, and afterward one of the highest standing in the army, had been stationed years before at Fort Riley, which was located near where now stands the flourishing town of Abilene, Kansas. This officer, in an open letter that was published in the newspapers at the time, denounced the directors of the "Kansas branch of the Union Pacific Railway" as swindlers, because they offered lands for sale in the vicinity of Fort Riley. He warned settlers not to purchase farms in that region, as the soil was so sterile as to be virtually a desert, and that their families would starve if they depended upon their own crops for sustenance. In less than a dozen years after the issuance of this manifesto a correspondent of the "New York Tribune," writing from Abilene, stated that he had seen from the cupola of the railway depot, extending as far as the eye could reach, thousands upon thousands of acres of luxuriant wheat, in gathering which were engaged hundreds of reaping-machines.

The Andes and the Rockies.—While the Andes have not a break, and loom up bluff quite close alongside the Pacific, from Panama to Patagonia, the Rockies on the fortieth degree of north latitude are nearly 1,000 miles east of that ocean, while the intervening space is interspersed with mountains, as the Sierra Nevada and the highlands of the Coast Range, and also with fertile valleys, through which flow streams the outcome of never-failing springs. The compact barrier of the Andes is in width about sixty miles, while the breadth of the Rockies is nearly four hundred ; the latter are not in a continuous, unbroken range, but they lie rather in blocks, between which are passes or depressions, that admit the passage of the winds freighted with warmth and moisture from off the Pacific, and also pave the way for railroads, thus rendering the east and west portions of the United States accessible to each other. Neither the Rockies nor the Alleghanies are so high as to prevent entirely the winds thus saturated with vapors passing over them, as do the Andes.

XXXV.

Origin of the Japan Current.—The interest in
this climate is enhanced by the fact that its mildness
and moisture are derived from off the Pacific, as those
of the eastern and middle portion of the Union are
from off the Atlantic. The two appear to meet and
blend between the meridians 106° and 108° west.
There comes up from that inexhaustible reservoir,
already mentioned, the Antarctic Ocean, a great
flow of water named from its discoverer the Humboldt Current. It is driven by the centrifugal force
induced by the rotary motion of the earth, and
dividing near Cape Horn one portion moves toward
the equator along the west side of South America,
thus corresponding with the other, which constitutes
the current on the west coast of Africa already
noticed as the origin of the Gulf Stream. As the
former approaches the wider part of the continent
and the equator, it becomes deflected toward the
northwest; that portion is known as the Peruvian
Current. This mass of waters at length turning
west constitutes on the line of the earth's swiftest

25

motion two flows direct across the Pacific—the south and the north equatorial currents, there being no land to obstruct or deflect their course as in the case of that of the Atlantic. The latter is the origin of the Kio Suo or Japan Current, for, while a portion passes on through the East Indies, a much larger one, turning northeast, at or near the Philippine Isles, begins its return course by flowing along the east coast of Asia. These waters flow on both sides of the Japan Isles, but by far the greater portion on the eastern, and being originally for a longer time and for a greater distance under the broiling sun of the tropics, they become warmer than those constituting the Gulf Stream. A United States survey found the highest temperature of this current to range from 84° to 86° Fahr. (*Lieutenant Brent, Perry's Japan Expedition.*) The volume of water is estimated to be three times greater than that of the Gulf Stream— that is, equal to 3,000 Mississippis at full flow—and its temperature being higher, the evaporation from its surface must be so much more in proportion ; but it also has a much greater surface to spread over and warm, since the North Pacific between the parallels of 36° and 40° on an average is about 5,000 statute miles wide, while that of the North Atlantic between the same lines is not quite one third as much. In addition, the warm temperature derived from the Gulf Stream, when its waters spread over the North Atlantic, is much modified by the cooling effect of the numerous icebergs that float down from the Arctic

Ocean, while, on the contrary, there *are none* in the North Pacific. Behring Straits are so narrow—being only about fifty miles—that icebergs can not pass, the waters being comparatively shallow, while the polar currents to transport them are wanting in force, when ranked with the power of those that flow south into the North Atlantic. From these facts we may estimate how much greater is the amount of moisture and warmth carried by the winds from off the Pacific than that borne off the Atlantic. Says Captain Maury : " The quantity of heat discharged over the Atlantic from the waters of the Gulf Stream in a winter's day would be sufficient to raise the whole column of atmosphere that rests upon France and the British Isles from the freezing-point to summer's heat." (*Physical Geography of the Sea, p. 54.*)

The Beneficial Effects.—The Japan Current with its immense volume of waters, saturated with a wealth of warmth and vapor, sweeps northward till it comes in contact with that great breakwater, the Aleutian Isles, extending in a southwest direction for nine hundred miles. This obstruction deflects its course toward the northeast and east ; meanwhile its advancing waters are also expanding over the Northern Pacific, and moving along the southern shores of Alaska, and round by British Columbia and Washington Territory to a point opposite the mouth of the Columbia River, where it turns toward the southeast, and passing at some distance from the coast, until it reaches the latitude of the peninsula of

Southern California, where its waters return into the original current going west, to be again warmed under a tropical sun, and again utilized to render the world a similar service; and thus they go on forever.

The Coast Climate.—The temperature of the North American coast that borders on the Pacific is remarkably mild, owing to the influence exerted by the Japan Current. At the Straits of Fuca, south of Vancouver's Island—48° 30′ north latitude—there is scarcely any ice, and flowers that it is necessary to house in the Middle States, during the winter, bloom here in the open air the year through. The climate is still more striking for its mildness farther up the coast, as at Sitka—57° north—in Alaska, the average temperature of the year is very nearly the same as that at Washington, D. C., 39° north, and also at Puget Sound—48° north—Washington Territory, the winters are almost as mild as those at Norfolk, Virginia, 37° north; and the same influence extends south along the coast beyond San Francisco.

The Region benefited.—The area in Northwestern America that is warmed and furnished with moisture from off the North Pacific is much greater than that of Northwestern Europe similarly affected from off the North Atlantic. The westerly winds off the former, having been saturated with warmth and vapor, penetrate the interior from the 42d to the 49th degree north latitude, for eight or nine hundred miles; even finding their way through the gaps in the Rocky Mountains and to their eastern slopes, where the vapor manifests itself by becoming visible in mists and clouds that descend in rains that cherish the native wild fruits and grasses, and, where the soil is brought under cultivation, rewards the labor of the husbandman. The same may be said of British Co- lumbia. From the 42d degree northward the Rocky Mountains decline in height, while their structure ap- pears to be more broken into blocks, between which are numerous depressions or gaps that furnish an easy passage for these westerly winds. The latter follow the valley of the Columbia River in a north- erly direction through the Cascades, and up beyond its head-streams near the 53d parallel in the British possessions, and only a short distance south of where the Canadian Pacific Railway passes through the Rocky Mountains from the plains in the valley of the Saskatchewan. The influence of these warm winds, passing through and over the mountains, affects the climate north of the British boundary of 49°, and so much that the temperature of the whole region, up to the 55th degree and east of the mount- ains, is greatly modified, especially during the sum- mer months. Though the winters are intensely cold, Nature seems to be in some way invigorated by her long rest, and the summer being so very warm, and the sun lingering so long, she is induced to put forth all her energies to produce and mature the wheat, the barley, the oats, and the grass. Dr. Rich- ardson, in his Arctic expedition, states that " wheat is raised with profit at Fort Liard, 60° north, in the valley of the Mackenzie River, at an elevation of 400 feet above the sea. This locality, however, being in the vicinity of the Rocky Mountains, is subject to summer frosts, and the grain does not ripen perfectly every year, though in favorable seasons it gives a good return." (*Climatology, p. 449.*) The great benefit accruing from this region to the portion of the Union bordering on the 49th boundary west of Lake Superior is that the cold that would sweep down from the Arctic Ocean is met and moderated by this barrier of warm temperature. When this current of warm air, thus saturated with vapor, in moving up the Columbia, reaches the mouth of the Shoshone or Snake River, a portion also passes up that stream, mainly in a southeasterly direction. Though this air is clear during the day, at night its vapor is condensed by the cooler temperature, and becomes visible in a fine mist, while the effect on vegetation is that of a very heavy dew, almost a shower, and the moisture penetrates the soil to such an extent that it needs no more rains after those of the spring are over. (*Smalley, p. 383.*)

Comparisons of Temperature.—On the South Saskatchewan River, 51° north, the average summer temperature is the same as that of New Haven, Connecticut 10° farther south. This mildness extends in a modified form still farther north, as we have just seen, and as also evidenced in the fact that buffalo winter in the woodlands along the rivers that are

tributary to Lake Athabasca, in a region lying between the parallels of 53d and 58th degrees north,
and obtain food from the native grasses of the previous summer. These grasses are sometimes covered
with snow, which the animals remove with their
hoofs. "Grassy plains like these necessarily imply
an adequate supply of rain, and there can be no doubt
that the correspondence with European plains in like
geographical position—those of Eastern Germany
and Russia—is quite complete in this respect. If a
difference exists, it is in favor of the American plains,
which have a greater proportion of surface-waters,
both as lakes and rivers." (*Climatology, p. 531.*)

At Fort Benton, on the upper Missouri, in Montana, 48° north latitude and 111° west longitude, and
at an altitude of 2,600 feet above the ocean, the average summer temperature, as ascertained during a
number of years, was 72° Fahr., and the average for
the year 48°; in New York Harbor (Fort Columbus)
the average summer heat is the same, while for the
year it is only three degrees warmer, though *seven*
degrees farther south. This is the more remarkable,
when we take into consideration the much greater
altitude of Fort Benton and indeed of that whole region, and that also the cooling effect of such elevation is overcome only by the heat which the Pacific
winds thus distribute. In the valley of the Red
River of the North, where the altitude is not so great
as at Fort Benton, at Pembina, Minnesota, on the
forty-ninth parallel, the average temperature of the

summer is 71°, one degree higher than at Harris-
burg, Pennsylvania, *nine* degrees farther south. The
higher temperature at Fort Benton may be accounted for, because it is 649 miles nearer the Pacific than Pembina. Another peculiar feature of that
region is the sudden and rapid opening of spring,
which usually extends a distance of 1,500 miles, from
St. Paul, Minnesota, northwesterly to the plains of
the Saskatchewan in the British possessions, within
ten or fifteen days. This marvelously rapid change
from the cold of winter is the earnest of the continuous warmth of summer, with its variable winds and
rains.

The Interior Climate.—Between the fortieth and
the forty-ninth parallels of north latitude, and within,
and east of the Rocky Mountains and west of the
Mississippi, the average summer temperature, as ascertained by observations at *seventeen* places located
in widely separated positions, is nearly 73° Fahr.,
while the average rainfall for the summer months,
and over the same territory, is fourteen inches, as obtained from statistics at *ten* stations equally separated.
(*Smithsonian Contributions, Vols. XXI and XVIII.*) As
already noted, the Rocky Mountains decline in height
from the forty-second parallel northward. In proof
of this may be cited the fact that the Northern Pacific Railway passes through them on a line lower
by some 3,300 feet than does the Central Pacific,
so that it has been characterized as "the valley
route across the continent." In the main it runs

through the central portion of the region just mentioned; though it is less incommoded with snows than the Central Pacific, yet its route for the most part is *four degrees* of parallel—or 276 statute miles—farther north. The mildness of the climate in this region gradually increases toward the Pacific, from a short distance west of the Mississippi, and is quite as uniform as the rainfall. Says Mr. E. V. Smalley: "In many portions of Dakota, Montana, and Northern Idaho, herds of cattle roam, and horses range out all winter, and keep in excellent condition on the nutritious grasses of the plains and valleys." (*Hist. Northern Pacific Railway, p. 174.*)

XXXVI.

THE much greater portion of the territory of the United States can be made available for the use of man, either as pasturage for herds of animals or for cultivation. In their native state, the plains, the valleys, and the hill-sides are for the most part fertile and susceptible of still further productiveness by the industry of man; and in addition, when the entire area of the Union is taken into consideration, the amount of rain is found to be abundant and uniform, while the temperature in its influence in maturing crops is equally efficient. Even the mountains amply compensate the Nation for the room they occupy, by furnishing minerals and moderating the climate.

The Value of the Valley.—In noticing more fully the Mississippi Valley we offer no apology, since it is a natural resource in its value to the American people pre-eminently beyond any other one of their possessions. Its vast area—more than half the entire Union, excluding Alaska—its position in the center of the Nation, and its accessibility to the other portions of the Union, so that wherever needed its products can be easily transported, unite in enhancing its national importance. It has the three conditions essential to make a country productive of grain, grasses, and fruits—a soil of itself fertile, rainfall and sunshine, and withal a temperature sufficiently warm to produce luxuriant and well-ripened crops. Being open at both ends, there is free ingress for the warm, moisture-loaded winds from the south to meet the drier and cooler ones from the north, by which the vapors of the former are condensed, so that the clouds thus produced may pour down blessings in copious rains. Nor do these blessings end here. The surface of the valley being in the greater portion moderately level or undulating, with here and there greater elevations, these rain-waters run off slowly, meanwhile penetrating the earth, to be stored therein, not only to nourish the crops, but to afford an abundance of pure water for man and beast. Everywhere, especially in the hilly portions, are never-failing springs and crystal brooks, while on the plains or level districts pure and sparkling water can be obtained in abundance by sinking wells.

The **Rapid Settlement.**—The unprecedented rapidity with which the central portion of this valley was settled, especially the prairie region lying within the peninsula between the Ohio and Mississippi Rivers, deserves notice. Rumor had told of that goodly meadow-land of grass and wild flowers, where the farmer, instead of laboring to clear off the trees and underbrush before he could bring the land under cultivation, had only to plow the prairie, put in his seed, and reap a crop in a few months. What an enhanced value accrued to this resource of land in its being at once available! Here was a remarkable opportunity for enterprise, and for which the people were then prepared. During the first quarter of this century commenced a migration from the older States of multitudes of young, energetic, and stalwart farmers, that almost swarmed over these plains, dotting them with farm-houses, while with comparatively easy labor they overturned the soil and consecrated the prairies to husbandry. Owing to the facilities of cultivating the land, these pioneers soon reached the Mississippi, and passed over to subdue the plains beyond.

Rainfall and Temperature.—Within this valley is now, according to the census of 1880, the center of the population of the Union—about eight miles southwest of Cincinnati—and in it may soon be the center of influence in the government of the Nation, as it certainly will continue to be its granary. (*Hist. of the American People, p. 1092.*) As a basis for the latter statement, the soil, so remarkably fertile by Nature, can be made for all time more and more productive by the skill and industry of man; should the conditions of Nature remain the same as they are to-day, the promise of seed-time and harvest will never fail in the Great Valley. The average annual rainfall within the valley is forty-two inches; that is

to say that if this water did not penetrate the earth
nor run off, it would be at the end of the year three
and a half feet deep. This knowledge has been ascertained by observations taken at forty-three widely
separated stations within its borders, ranging north
and south from 30° to 45°, a distance of more than
one thousand miles, and from distant points east and
west. (*Smithsonian Contributions, etc., Vol. XVIII.*)
The ratio of the rainfall during the summer months,
when it is specially needed, is greater than during
any other season, averaging four inches each month,
thus furnishing moisture to promote the growth of
cereal crops and grasses, native and cultivated.
Then, again, the necessary sunshine and heat are provided to mature and fully ripen these crops; the
high temperature thus utilizing this large rainfall.
As ascertained from observations taken at thirty-
four stations covering the same parallels of latitude
and at equal distance east and west, the average temperature of the valley is found to be 75° Fahr. during the three summer months, and the average temperature of the entire year 51°. (*Tables of Temperature—Climatology.*)

The Atlantic and Pacific Slopes' Rainfall.—The
annual rainfall on the Atlantic slope averages forty-
three inches not far from the coast, and that of the
summer months is nearly thirteen. Yet near Cape
Hatteras it is seventy-eight inches, because of the
blending at that point of a cold northeast current
with the warm water of the Gulf Stream.

On the Pacific coast of the United States we find a difference in one respect: in that region the most profuse rains are in the winter and spring, by which means the soil is prepared to afford moisture to the growing crops, and also sufficient to sustain them during the months when they are ripening under an almost cloudless sky. As shown by observations taken at ten different stations, the average ratio of rainfall for the winter alone was eighteen inches, or four and a half for each month, while for the entire year it was only thirty-six inches—six inches less than that in the valley of the Mississippi. (*Smithsonian Contributions, Vol. XVIII.*)

Average Rainfall in Europe compared.—The feature of the winds saturated with vapors penetrating into the interior, in the Northwest, and also in the valley, so far away from their source, is very striking when compared with the short distance they are borne into the interior of Europe. At St. Petersburg the annual amount of rainfall is not quite eighteen inches, while its distance from the ocean is only 525 statute miles; and in the Crimea, in the vicinity of the Black Sea, it is only fifteen inches; and at Orel, 484 miles north of that sea, it is twenty-five inches. At Berlin, Prussia, the rainfall is twenty-four inches, and in Paris twenty-three. In the British Isles, at Liverpool it is thirty-four inches; Aberdeen, Scotland, twenty-nine; north of Ireland, thirty-six; and, though the fogs of London are proverbial, the annual amount of rainfall in that city is only twenty-one inches. The average rainfall in the United Kingdom of Great Britain and Ireland varies more than on the Continent; it sometimes attains thirty-five inches. A more comprehensive statement gives the annual average rainfall of Europe at twenty-four inches, as ascertained by observations made in seventeen cities that are widely separated. It is interesting to note in this connection that *twenty-four inches* is the annual rainfall at Fort Riley, near Abilene, Kansas, and also that here the average summer temperature is 77° Fahr. This town is on or near the 39th degree of north latitude, at its intersection with the 97th meridian west, and here—excluding Alaska —is the territorial center of the United States (*Hist. of the American People, p. 1092*)—the latter position being 1,404 statute miles from the Atlantic and the same from the Pacific; while on the south, from the west end of the Gulf of Mexico, it is 770, and on the north about the same from the British possessions.

The **Center of Territory.**—The reader will notice the great distance of this center of territory from both oceans, and also that the average annual rainfall at that point is equal to the similar average of *all* Europe. This result shows that the winds that blow north over the valley of the Mississippi are more deeply saturated with vapor off the Atlantic Ocean current than are the winds that float east over Europe from off the northern portion of the Gulf Stream; the waters of the latter being so much cooler, the evaporation is correspondingly less. We may note in this connection that the average annual rainfall of the United States—excluding Alaska—is twenty-nine inches, as we have seen that of Europe is twenty-four. (*Appletons' Physical Geography, p. 131.*)

The Soil of the High Plains.—We have already noticed the immensity and the various kinds of mineral wealth that are stored in the earth within the territory of the United States, and yet the benefits which the people derive from the soil of their country far transcends in value the treasures found beneath its surface. Taken as a whole, the fertility of the soil of the Union is the most remarkable on the globe, and its productions are correspondingly great. Within the territory of the United States, as already noticed, are no great deserts in the sense of purely barren wastes, though there are two or three sterile districts which are comparatively small; but none of these compare in size or partake of the dismal, sterile features of the deserts of Africa and Central Asia. Instead, they are mere specks of barrenness, when compared with the entire mass of the fertile, arable, and pasture lands of the Union.

Mistakes of Explorers and Others.—Early explorers, tourists, and even army officers, were mistaken, especially in their estimate of the productive qualities or fertility of the portions of the elevated plains which were near and east of the Rocky Mountains. The sage-brush plains were pronounced barren and worthless, and the entire region was designated on the map as the Great American Desert; but now the same space on the map contains the names of prosperous villages, in the vicinity of which are fruitful and cultivated farms.

More recently the "bad-lands" of Dakota were described as utterly worthless; but time has shown that "they are full of rich plateaus, and afford the best of shelter for stock. They were the favorite · haunts for wild game, and are now (1886) occupied by numerous and consecutive herds of cattle, that' have been driven to them during the past five years. They have proved to be anything but 'bad' to the stock-growers of the plains. Buffalo and bunch grass cover almost every inch of the ground." (*Dakota Handbook, p. 23.*) "The fertile prairies and valleys of Western Dakota terminate near the western boundary of the Territory, in that singular and picturesque region known as the 'bad-lands' of the early maps. This region resembles no other district of country in the world. Yet everywhere in its valleys, save on the faces of the steeper buttes—a conical hill that rises out of the plain, or rather, in the course of many centuries, the surrounding earth was washed away from them—the grass grows luxuriantly, covering even the high plateaus five hundred feet above the valleys, and here herds and flocks find pasture the year round." (*E. V. Smalley, Hist., etc., pp. 338, 339.*) "The autumn is the most agreeable season; in summer it is as hot as in southern latitudes; the winter begins in November and lasts

26

until April, and most winter days are clear, bright, and still."

In Salt Lake Basin, stock generally winters without prepared fodder, but thrives on the range the year round. "The seed of the bunch-grass has remarkable fattening properties. It grows in bunches in apparently the most barren places; early in the season it cures, and still standing retains all its nutriment, and, being hard to cover with snow, it is within the reach of stock." (*Resources of Utah, p. 19.*)

"Although Arizona has been represented as a barren, sandy waste, it can show as fine a growth of rich and succulent native grasses as any section of the Southwest, and its capabilities as a stock-growing region are almost limitless. The cattle fed on this grass keep fat winter and summer, and their beef is unequaled in flavor. The fattening qualities of these grasses are almost beyond belief." (*Resources of Arizona, pp. 63, 258.*) This Territory has 60,000 square miles of pasturage on native grasses. New Mexico has grazing-lands in great extent, the grasses being the same in kind and qualities as those of Arizona; with these her plains and mountain-slopes are covered. This grass does not flourish on damp or clay soil, and hence it is not found in the river-bottoms or valleys, but thrives best amid sand and gravel, and in perfection on dry sandy plains and rocky hill-slopes. No kind of stock in New Mexico is ever required to be winter-fed or sheltered. These immense pasture-lands extend into Texas and Kansas, Indian Territory, and Western Arkansas.

The Native Grass Resource.—The purely natural resources of the native grasses on these high plains and mountain-slopes have only within recent years been appreciated. They never fail, as they are fitted by Nature to the conditions under which they exist; they need not the care of man, but flourish where the ordinary cultivated grasses would wither and die. They are specially adapted to the nourishing and raising of herds of cattle and flocks of sheep, and this will continue as long as the present laws of Nature prevail in that vast region. Throughout this immense pasture-range, on high plains and hills, are also often found valleys and lower lands with rich alluvial soil, but almost unproductive because of the lack of rain or moisture. These districts the farmers easily convert into fruitful fields and gardens by means of irrigation. (See Section XXXVII.)

The Discovery.—It was known that, during the winter, the buffalo and the deer fed on cured grass or wild hay; but could domestic cattle be thus carried through that season? This problem was accidentally solved in the winter of 1864 and 1865. A teamster was hauling supplies with oxen to a United States fort in Utah, when he was overtaken by an early snow-storm on the Laramie Plains. He was compelled to halt for the winter; when his feed was exhausted he turned his oxen out, as he supposed, to die of starvation. The latter did not stray far away, but lingered in the vicinity of the camp, and when spring opened they were found to be in a much better condition than when they were turned loose. The winds had here and there laid bare the cured buffalo-grass, and upon this the oxen had fed for nearly four months. The important discovery that cattle could winter well on wild hay paved the way for establishing cattle-ranges throughout that region and farther south. This has since grown into an immense business, and, in cheapening one important class of food, has conferred great benefits, not only upon our own people, but upon those of Europe, who have become participants in the advantages derived from these cattle-ranges.

XXXVII.

THIS mode of nourishing crops has been in use from the earliest antiquity—on the plains of Assyria, in China, in Egypt, in Italy, while Pizarro found in Peru a perfect system of irrigation. It is only within comparatively recent years that the process has been found necessary within the domain of the Union.

The United States are remarkably blessed with water—Nature's great fertilizer—in their copious rainfall, not only on the Atlantic slope and in the valley of the Mississippi, but in the greater portion of the territory west from the Sierra Nevada to the Pacific. There is, however, a region in the Union, whose length is about 800 miles, with a width ranging from 100 to 150, within which are large isolated areas where the rainfall is not sufficient to nourish properly the growing grains and vegetables. This comparatively dry region extends from the middle of Idaho and Wyoming, across Utah, Colorado, New Mexico, Arizona, and southwest, through Nevada, into Southern California. Through the middle portion of this region run the Rocky Mountains, and at quite a distance from their eastern base the western flank of the rainfall that supplies the Mississippi Valley becomes partially exhausted. On the other hand the vapors that rise from off the Pacific, and are carried east by the winds, in passing over the Sierra and Cascade Mountains become partially chilled, and deposit on them much of their vapor in masses of snow or moisture, while the remainder passing on descends in snow amid the Rockies; in consequence, the rainfall in the valleys of the Great Basin is much diminished. The native grasses, however, not requiring nearly so much moisture for their sustenance, live and flourish on the side-hills and on the upper plains, throughout this entire region, and thus furnish abundant pasturage for stock.

The Deficiency supplied.—Throughout this area, thus lacking sufficient rain to perfect the growth of domestic vegetation, the farmers resort to irrigation to supply the deficiency; to accomplish which they avail themselves of the inexhaustible stores of snow upon the mountains, which in due season melts under the rays of the summer sun, and the water trickles down in streamlets, which are often picked up one by one, and concentrated in canals, and thus carried where the life-giving water can be used in nourishing the roots of plants and of grain. It is a remarkable provision, in the natural economy of the United States, that the fertile valleys and their adjoining slopes within this area, which, for agricultural purposes, are almost valueless, because of the want of water, have within reach a supply of the latter that is absolutely inexhaustible. To the farmer living in regions where abundant rains prevail, the labor of irrigating his fields would no doubt appear toilsome; but the latter mode has its compensations, for he who irrigates is not left to the mercy of droughts nor of chance rains, nor is he delayed by untimely wet days during the growing season, nor by storms when he is harvesting his crops.

The Canals.—The mode of bringing the water from these mountain reservoirs has become more and more systematized, from the private canal of the individual farmer of only a few hundred yards in length, to that of the stock companies, ranging in length from a few miles to sixty or eighty. These companies, utilizing mountain-streams, build canals and furnish water to the farmers at a fair rate, and in return receive fair dividends from their investments. On the other hand, the farmer pays for the water he uses, but is himself amply compensated for his out-lay in the enhanced produce of his fields. He easily learns how to use it to the best advantage, such labor being almost nothing; while the soil of that region is very easily permeated by the water, and the re-sult is abundant crops. This mode of supplying the necessary water for raising crops prevails to a cer-tain extent throughout that entire section of coun-try, and "hundreds of thousands of acres, that were barren wastes a few years ago, have been brought into a high state of cultivation by means of irrigation."

The Effects produced.—Singular effects sometimes occur from irrigating certain soils, as, for instance, in Colorado: a farm may be taken up amid the sage-brush, and, when cultivated by means of irrigation, the water decomposes the alkali of the soil, and converts it into a fertilizer. Almost everywhere in that section can be seen the productive field side by side with the original sage-brush. Irrigated lands, as a general rule, when properly cultivated, do not become poor and sterile, but continue to produce fine crops of wheat, oats, barley, potatoes, and garden vegetables. The slope lying directly along the eastern base of the Rocky Mountains was designated on the earlier maps as the " Great American Desert"—a barren waste, but now known to be susceptible of cultivation, and under certain conditions producing remunerative crops.

In Colorado, large canals have been constructed, one of which is eighty-two miles long, and has capacity for bringing water sufficient to irrigate 65,000 acres. Says Governor Elbert, in an address: " Crops with a regular supply of water, from the first impulse of spring until they ripen for the harvest, are developed to their fullest capabilities, and in the greatest perfection." This statement has been verified from year to year by the average yield of crops in the portions of Colorado where agriculture is exclusively carried on by irrigation. The modes and facilities for making the system perfect are increasing from time to time, and this valuable *resource* of melted snows is more and more appreciated.

In Wyoming, irrigation has been extensively introduced, there being nearly ninety stock companies engaged in supplying water for that purpose to the farmers on the plains in the southern portion of the Territory. These companies, in addition to utilizing the mountain-streams, great and small, have often recourse to artesian wells. The individual capital ranges from the lowest on record, $300, up to the largest, $1,000,000 in two instances. The main canal belonging to one of the latter, including its branches, is 200 miles long. From the report of the Governor (1885, p. 77), we learn that the development of the agricultural industry of the Territory has been very great. The grain produced by this method is better, weighs more to the bushel, than that obtained by the ordinary process, as the soil—a fine, sandy loam—is well adapted to become fruitful by irrigation. The same mode of farming, and to a large extent, prevails also in Idaho, and in both instances has been eminently successful, though the system is only in its infancy.

Southern California is not so happy in obtaining water for irrigating purposes from streams derived from the melting snows in the mountains, as snow is quite limited in that latitude, but it depends upon other natural streams, springs, and artesian wells. The latter are rapidly increasing in number, there being about 600 in Los Angeles County alone. These furnish an immense amount of water, which is usually pumped by means of windmills. Crops of every kind—fruits and grain—are very large in that portion of the State.

In Nevada, irrigation is greatly needed to make a large proportion of its soil productive, and the time seems not far distant when State legislation will adopt measures by which the system can be universally introduced, so as to benefit equally all classes of citizens. (*Land Register's Report, 1885, p. 6.*)

Suggestions.—This subject has attracted the attention of the National Government, and the Secretary of the Interior says in his report, December 6, 1886 : " It is urged that a system of irrigation, which is necessary to render inhabitable immense areas of fertile lands in the Territories, is impossible to small holders. The sinking of artesian wells, the construction of extensive reservoirs and irrigating canals many miles in length, must be done by accumulated capital, and that therefore, in the development of the Territories, large masses of land must be brought in this respect under a single management." The Secretary suggests that " Government itself should undertake preliminary scientific surveys and investigations," and, perhaps, conduct a system of public works to promote irrigation, and thus aid in establishing within the Territories a numerous and wealth-producing population.

XXXVIII.

WHEAT—that princess of cereals—is cultivated over a greater extent of territory in the United States than any other grain, except one—Indian corn. The area producing it is more than half that of the Union; it extends from the Atlantic to the Pacific across the middle portion of the continent in one continuous belt, having a width of several hundred miles, and, in addition beyond these east and west lines, it reaches far toward the south along the highlands taking in the foot-hills and slopes on both sides of the Alleghanies; then, sweeping round toward the southwest, flourishes on the elevated region constituting the divide between the watershed of the Gulf and that of the valley of the Ohio; thence, skipping over the lowlands bordering the lower portion of the Mississippi Valley, it resumes its sway in the elevated region of the Ozark Mountains, and south along their line into Northern Texas, and still moving westward to the Pacific slope, where in every part it finds a congenial home—from Southern California across Oregon, Washington Territory, and far into British Columbia. Nature has furnished such a soil and climate in some portions of California that two crops a year—wheat or barley, and Indian corn—can be produced from the same ground by means of judicious irrigation.

The Northern Wheat-Belt.—The main wheat-belt includes New England and Central New York, though its cultivation in these sections has been partially suspended, because of the much greater facilities afforded in the West and Northwest, by the more fertile soil and open plains; the latter being susceptible of easier cultivation, and also of producing larger crops and at less expense. Directly west of New York comes in a portion of Canada, that fine wheat-growing region, the peninsula between Lakes Ontario and Huron; but west of the Mississippi the wheat area extends still farther north, along the valley of the Red River till it reaches Lake Winnipeg, and thence expands over the plains through which flows the Saskatchewan, in north latitude ranging from 50° to 53°.

Wheat's Adaptability.—Wheat has the power of adapting itself to the conditions of elevated regions, and likewise it appears to flourish better and produce with more vigor in those sections where the climate has almost the extremes of heat and cold. The intense winter cold of this entire region just mentioned has its compensations, as the earth often contracts because of the cold, and portions thus separating produce cracks or open spaces two or three inches wide and extending in depth several feet. These openings on the melting of the snow become filled with water, and, as the frost leaves the ground, the soil crumbles and shuts it in these little chasms. The latter become within the earth innumerable though small reservoirs, whose waters in the warm and dry season nourish the roots of the grains and grasses. Owing to this provision of Nature, often excellent crops of wheat have been raised in the valley of the Red River of the North, without any rain to wet the ground after the wheat had been sown in the early spring.

The Effect of Sunlight.—It is remarkable that the wheat of what we usually term the " Northwest " is more than ordinarily perfect; the kernel is said to be harder and more compact than that grown farther south, and that this applies, with equal if not greater force, to the wheat produced on the plains in the vicinity of Lake Winnipeg. Prof. Schubeler, of Christiania, Norway, after studying and experimenting on the subject, has shown conclusively that long and continuous sunlight has a marked effect on vegetation in perfecting the peculiar quality, whatever that may be, of the grain or plant. This result may be produced by the actinism or some other element contained in the sun's rays, and which have the same influence everywhere, and apparently independent of heat, though the latter promotes the general growth of vegetation. In accordance with Prof. Schubeler's experiments, the actinism of the sun's rays appears as effective in its action in the north as it does in the south. May not this be the secret of the very excellent wheat-kernel produced in this northern region? Here is a problem for further scientific investigation.

The Long **Continuance of Sunlight.**—An additional reason may be given to account for this beneficial effect produced on the wheat, while growing in this northern region—namely, the longer continuance of the sun's rays. For illustration, taking the average rate of the rotary motion of the earth on its surface within the belt included between seven degrees of north latitude—the forty-sixth and fifty-second inclusive—the rate is found to be only *three fifths* of the surface rate at the equator. From these conditions it follows that, for instance, on the 21st of June, when the length of the day from sunrise to sunset is *fifteen hours*, the sun shines on an area—of any assumed size, say a square statute mile—at the equator for *six hours*, but on an *equal area* within this belt, just mentioned, it shines *nine hours*, owing to the slower motion of the earth's surface, and in consequence the efficiency of the actinism of the sun's rays within this belt is increased *one half more*. It is found consistently with the above that—dispensing with small fractions—the circumference of the earth and the length of a degree on the forty-ninth parallel are only *three fifths* respectively of that on the equator.

Wheat in the Great Valley.—By far our greatest and most connected wheat-producing area is the valley of the Mississippi; this includes the valleys of all its tributaries with the adjacent side-hills and plateaus. Within this area wheat flourishes, except in the lowlands in the extreme southern portion. This granary of the Union, and to a certain extent storehouse of Europe, deserves a passing notice. Throughout its length and breadth there is in the main a depth of alluvial soil whose fertility to those who have never seen it seems almost incredible. In the bottoms along the river itself, and in those bordering on its tributaries, the soil is deep, often ranging from two to ten feet, and in extreme cases even twenty or more. The hillsides within the valley, nearly all of which have a limestone soil, produce abundant crops. The wealth of this natural resource can not be strictly estimated at a mere money value; its province is to furnish sustenance and other blessings for the people of to-day, and to promote, indefinitely, the well-being of coming generations. Meanwhile, its intrinsic value will, no doubt, continue to be enhanced by means of the intelligent industry and skill of the future agriculturist. This area is so extensive, the climate so diversified, the sunshine and rainfall so assured, that though droughts may sometimes occur in different sections, yet, humanly speaking, there never will be a deficiency of crops so great as to cause a famine throughout the land.

Large or **Small Farms.**—It is apprehended, in theory at least, that the soil of the numerous very large wheat-farms in the Northwest, which to-day are so extensively cultivated, must in the end become exhausted from the lack of proper care. This theory assumes that farming on so large a scale will not replenish the soil with the plant-food taken from it by the crops, and in consequence in time these lands will become unfruitful. Such very large farms are, from the nature of the case, exceptional, and they will not remain so for a great number of years, but will, in accordance with American ideas, be broken up and pass into the possession of small landholders. The latter will be more likely to cultivate them carefully, so that the soil will hold its own, and indeed become more fertile from year to year; meanwhile the aggregate of the crops will be larger than when under the management of a single owner; in addition, the land will be occupied by a much larger population, and confer benefits upon a greater number of people.

The Maize or Indian-Corn Area.—Maize is the only one of our cereals that is indigenous to the United States. When the colonists obtained it from the Indians, they called it *corn*, the English term to express simply grain of whatever kind; afterward, to make the designation more definite, the Americans, instead of retaining the Indian name *maize*, prefixed to *corn* the word *Indian*.

The maize area is even larger than that of the wheat, since it is not only continuous with the latter, but extends farther south away beyond its limits, and luxuriantly flourishes side by side with the cotton, the rice, and the sugar-cane. In the cooler northern portions of the Union it does not grow so freely nor produce so abundantly, as in the warmer middle and southern. This area, commencing in Eastern Maine, and passing west, includes New England and Central New York, and also extends along the curve of the Atlantic slope, running far back from the coast and taking in the foot-hills of the Alleghanies as well as the lowlands nearer the ocean, until it reaches Southern Florida ; thence westward, including the southern end and slopes of the same mountains, it covers the region north of the Gulf to the Mississippi, and beyond that river, over the plains of Texas and across New Mexico and Arizona to the Pacific slope, up which it passes from South-ern California to British Columbia. But the favor-ite home of the Indian corn is the area drained by the Mississippi and its tributaries—stretching from the Rocky Mountains to the Alleghanies ; here are the fertile bottom-lands and the rich hillsides in which it delights, and the copious rainfall and the warm sunshine that make it rejoice.

Extent of Crops compared.—A brief comparison may give a partial view of the immense value of this crop—our great indigenous grain, which next to rice supplies food directly or indirectly to the largest number of the human race. In 1880 the corn-crop was 1,754,861,535 bushels. This was more than two and a half times the next greatest crop—the wheat— the corn having 816,163,773 bushels to spare. It had

27

767,647,119 bushels more than all our other crops of grain combined—the wheat, the rye, the barley, the oats, and the buckwheat. It is quite probable that this ratio of production, or nearly so, will ever continue. The fattening properties of this grain are very great, and as such it is used for domestic animals, especially swine; in relation to the latter, in their being prepared for food as pork, corn is utilized to an enormous extent in the valley of the Mississippi. In the form of pork and beef, which are prepared in several ways, corn thus indirectly reaches the people of the whole country, while it is sent abroad in immense quantities. In addition, we have no grain that is so susceptible of being prepared in various ways for human food, nor one that is more nutritious and palatable.

Indian corn is far more used directly as human food in the South than in the North, yet it supplies an immense amount to the latter in two forms: the one directly as bread, and the other indirectly as in beef and pork, which the corn mostly fattens. The colored people have a knack of preparing this nutritious grain in a manner that renders it exceedingly palatable, and thus it supplies nearly all their bread, while it enters so largely into pork, almost their only meat.

Oats, Buckwheat, Barley, Rye.—The area that produces oats is very extensive, as they are cultivated in every State and Territory. The buckwheat area is quite limited, and its product is much lower than any other grain ; in some States it is not cultivated at all. Barley is raised in all the States and Territories, except Louisiana. Rye loves the middle and north-middle belt of the country. The amount produced in bushels of these grains in 1880, according to the census, is in the following order, corn being the first: Indian corn, wheat, oats, barley, rye, and buckwheat.

Rice-Culture and Area.—In 1698 the captain of a ship that put into Charleston harbor, in distress, presented to Archdale, the Governor of the Colony of South Carolina, a bag of rice, which he had brought from the Island of Madagascar. This rice was distributed as seed among the planters, in order to ascertain if the conditions of the soil and climate would promote its cultivation. The venture was successful; the grain produced was exceptionally perfect, and in consequence the industry advanced rapidly, and ere long " Carolina rice was celebrated far and wide as the best in the world." That reputation it has kept ever since, and " Carolina rice heads the list in the quotations of that article in all the markets of the world." (*Handbook of South Carolina, p. 12.*) The rice-producing area, with occasional breaks intervening, extends from near the Virginia line along the South Atlantic, and round on the north shore of the Gulf of Mexico into Texas. It runs in places quite back into the upper country, as well as occupies the districts on the mainland that can be flowed with water, and the adjacent islands that are suitable for its culture. There are two methods in cultivating rice—the dry and the wet. The former is pursued in uplands and on grounds not susceptible of irrigation ; the dry culture yields from fifteen to fifty bushels to the acre, the wet from forty to eighty. Of the States, portions of which are included in this area, those that are prominent in the cultivation of rice ,are South Carolina, Georgia, Louisiana, and North Carolina, in the order in which they are named.

Sugar-producing Area.—Our native sugar, for the most part, is derived from three sources: the sugar-maple trees of the more northern portion of the Union, and the cultivated sugar-cane of the southern, and from the sugar-beet. The cane, being of tropical origin, can be raised quite successfully in the region bordering on the tropics, or the sub-tropical. It is extensively cultivated in Louisiana, where it flourishes in the rich alluvial soil, but some authorities on the subject assert that it grows more vigorously in the south-middle portion of Florida than in the former State, and in proof of which they adduce the fact that in Florida it matures sufficiently to come out in tassel. In the latter State appears a wide field for its cultivation, but which as yet is only sparingly developed. Georgia, also, puts in a claim for her tier of counties bordering on Florida as being a valuable sugar-cane area, and Alabama for her territory near the Gulf.

In addition, sugar is also obtained from a certain kind of beet, which is produced in large quantities in

California, and the indications are that it can be cultivated to advantage in almost every portion of the valley of the Mississippi. Nearly the same may be predicated of sorghum, which has been introduced from China. This plant can be grown in the middle portion of the Union, but to what extent it will be utilized in the production of sirup or sugar remains a problem. We only know at present that the area over which it can be successfully cultivated is quite extensive.

XXXIX.

FIBERS—MISCELLANIES.

Cotton Area.—This area extends over the entire southern portion of the United States, and is celebrated not only for its large production of cotton to the acre, but for the excellence of the fiber; these two elements combined give that staple of the Union almost the monopoly of the commercial world. The value of this crop, because of its ready sale in Europe, has been estimated beyond its intrinsic worth, as it is in that respect the fourth in order; the Indian corn, the wheat, and the hay being each of more value to the people at large. To define more strictly this immense area of more than 100,000 square miles, we may state that it takes in the South Atlantic slope from the coast to the foot-hills of the Alleghanies, and at a similar distance from their southern end includes the territory between them and the north shore of the Gulf, and also a portion of Tennessee west to the Great River, crossing which it pervades a large portion of Texas, and also claims a part of Arkansas, and some districts in Arizona. Of this extensive area only a comparatively limited portion is annually planted in cotton ; the remainder is in trust for the future.

The Grades of Cotton.—There are two classes of cotton produced in the Union—the upland and the sea-island ; the short and the long staple. The upland is far in advance of the sea-island in respect to quantity, but the latter is far superior in relation to fineness of quality, yet both grades are essential in order to comply with the requirements of the American people. The sea-island or long-staple cotton is famous the civilized world over for the silky fineness of its fiber. From it is made the finest thread for sewing, and the muslins of Switzerland and France, so celebrated for their delicacy of texture.

Along the South Atlantic, sea-island or long-staple cotton is produced near the coast and on the adjacent islands within portions of three States—South Carolina, Georgia, and Florida. That area or region is capable of producing this fine and silky cotton to an almost unlimited degree.

Other Fibers ; Hemp, Jute, Flax, etc.—The area producing hemp is quite extensive, though it is principally cultivated in portions of the States of Kentucky and Missouri, the latter rivaling the former. It requires a rich soil, and in Missouri the term " hemp-land " is deemed a compliment to the fertility of the land thus designated. (*Missouri Handbook, p. 76.*) Hemp can be raised to advantage in several other States in the valley of the Mississippi when there is for it sufficient demand. *Jute*, a coarse kind of hemp, has been introduced from India into the South, especially in Florida and Louisiana; in each State it appears to find a congenial climate and soil. Plants that produce fiber appear to flourish in the southern portions of these States. Ramie was introduced from Southern Asia, and its cultivation has been successful as far north as New Jersey. From it is made the noted "grass-cloth" of the East; the fiber has a gloss similar to silk, and it is stronger than hemp. The Sisal hemp is indigenous to Florida, and several other varieties of that plant can be cultivated to advantage. The leaves of the *palmetto* are also very fibrous, but of a coarse texture, so that when prepared they are used in making brushes, brooms, and the unbroken leaves also make good roofing. When young and tender the palmetto-fiber furnishes materials for bonnets, hats, etc., as well as for the upholsterer, the cordage-maker, and pulp for fine grades of paper.

The *flax*-producing area extends across the north-middle portion of the Union; it rejoices in a moderately temperate climate and fresh soil. In the eastern part of this area its culture has been much diminished within recent years, the labor in preparing the fiber for the loom being so great that cotton to a very large extent has superseded its use. In the Northern States and Territories west of Lake Superior flax is cultivated almost entirely for the seed, which is extensively used in obtaining a valuable oil, known as linseed, and used in painting and for other

p̤urposes. Efforts are in progress, especially in Minnesota, to utilize the fiber by means of machinery.
(*Illustrated Minnesota, p. 23.*)

The northern part of the Pacific slope has a climate and soil peculiarly favorable to the production
of flax. This may be safely predicated of the western portion of Washington Territory and Oregon,
within the region between the Coast Range and the
Cascades. For illustration: the climate of the valley
of the Willamette—the "garden of Oregon"—is similar to that of Ireland, as there only rain and mist
prevail, and snow is scarcely known, while the temperature is equally mild. In this valley and vicinity
flax grows luxuriantly, and perfects its fiber to such
a degree that enthusiastic patriots predict that, in
time, this valley will have linen-factories which will
rival those of Richardson, on the banks of the Liffey.
In this region the "*shamrock*" grows spontaneously.
(*Oregon as it Is, p. 61.*)

The Tobacco Area.—Tobacco, being indigenous
to the country, its area virtually covers the whole of
the arable lands of the Union, except in the more
northern portions, though its favorite habitat is in the
south-middle region. It is a sensitive plant in respect to the elements in the soil that afford it food,
though it appears to flourish, without exception, in
fresh ground or that recently occupied by native
trees or underbrush; it is also susceptible to the nutriment derived from suitable fertilizers. Owing to
this fastidiousness in selecting plant-food, its cultiva-

tion is limited to districts in the area where that special class of food abounds in the soil; these districts, however, are often widely separated. In consequence of this peculiar trait, the texture of the tobacco-leaf and its flavor are found to depend very much upon the character of the soil in which it is grown.

Potato Area.—The area of the sweet-potato covers the south-middle and southern portion of the United States, where it is indigenous, as it was found in that region by the early English colonists, and where it flourishes in different varieties and produces in great profusion. The *white potato* was originally obtained from the highlands of Peru, where it grew in a wild state as a comparatively small tuber. It is known also as the Irish potato, because it is cultivated so extensively in Ireland. The present size and good qualities of this valuable esculent, and its numerous varieties, are the outgrowth of its careful cultivation during two centuries. Its main area, though it somewhat overlaps that of the sweet-potato, is in the middle and northern portion of the Union from the Atlantic to the Pacific.

The Peanut.—We must not overlook the humble peanut, which also grows in the ground, and used to be called the *ground-nut*. It is a native of Africa, and was brought to this country somewhere about 1850. This vegetable is grown in the light, sandy soils of Virginia, North Carolina, and Tennessee, and is making its way gradually toward the West, and appears

to have become thoroughly acclimated. Its production is increasing from year to year, to meet the corresponding demand, while the fruit itself is an improvement on the original African nut. This latter
phase renders the crop still more valuable. The
vine and leaves of this plant somewhat resemble
clover; they are cultivated like sweet-potatoes, and
when ripe the vines are pulled up, to the roots of
which the nuts adhere very tenaciously, and require
to be removed by the hand.

Hop-Production.—In Central New York is an extensive district where the hop grows luxuriantly,
and where its cultivation is a prominent agricultural
industry. The conditions of soil required for the
successful raising of this crop appear to be within
limited districts in the State. The quality of the hops
produced is, however, of the best grade. A remarkable instance of this adaptation of soil occurs also
in Washington Territory, in the rich alluvial, sandy
deposits of the river-bottom lands adjacent to Puget
Sound. The area brought under cultivation in hops
is rapidly extending, inasmuch as the climate as well
as the soil in the bottom-lands and valleys of the
Territory appear to be peculiarly adapted for raising this crop. The amount produced to the acre is
large—the average yield being 1,500 pounds—and
the quality of the crop compares favorably with that
of New York, while in the latter respect it is improving from year to year. The increase in the
acreage planted, and the amount produced in the

Territory, has been very rapid and extremely variable; for instance, in 1880 the crop was 5,000 bales; in 1884, nearly 22,000; while in 1885 it was only 13,000, the latter falling off being in consequence of the overproduction of the previous year. One half the crop goes to England.

Broom-Corn.—While as a general rule farmers throughout the country can raise broom-corn to supply domestic wants, in portions of Central New York it is cultivated as a farm-crop, and to such an extent as to supply brooms for the home market and also for Canada.

Peppermint.—The entire crop of this article in the United States amounts annually to about thirty-five short tons (2,000 pounds), of which nearly one half is exported. Two thirds of the peppermint-oil produced in the country is manufactured in Central New York, where the same proportion of the crop is raised; the remaining one third is made in the State of Michigan.

XL.

TIMBER is a natural resource, since the native forests are not the result of man's labor, but the product of Nature. The territory of the United States, compared with that of other countries, is remarkably well timbered. Said an intelligent gentleman, when speaking of a recent tour round the world: "From the time I left the United States till my return I did not see a group of trees that deserved the name of forest." Though the "prairie States" were originally devoid of timber, we have seen (p. 379) that in the main these fertile plains, thus destitute, became in consequence a far greater blessing to the nation at large, and especially to the people who migrated thither, than if they had been covered with dense forests. If in the cultivated sections of the country there is now a scarcity of timber, it has been the result of a lack of foresight in those who owned the land, in their not preserving more carefully portions of the primitive forests. Notwithstanding this neglect, an intelligent survey of those portions of the country once covered by forests, but

now for the greater part under cultivation, will discover almost everywhere intervening districts in
which still exist large remnants of the original forest, or are instead covered by a second growth—an
earnest of the future. The remedy for the scarcity
of ordinary timber in time to come is in the hands
of the people themselves.

The **Extent of Forests.**—In addition to the scattered remnants to which allusion has been made, we
may take a glance at the extent of our other forests.
In New England we have the noble pine-woods of
Maine, and those in the mountains of New Hampshire and Vermont; the Adirondacks and Catskills
of New York, with the highlands of Northern New
Jersey; while the Alleghanies, far superior to all
these, extend from the Canada border nearly 1,000
miles to their southern termination in Georgia and
Alabama, with their foot-hills jutting on their slopes,
east and west, in an average width in that direction
of nearly 100 miles; these hills with the main central
ridge are clothed to their very tops with noble trees.
East and southeast of the Alleghanies come the immense forests of North Carolina, South Carolina, and
Georgia, while on the north of the Gulf, including
Florida, they extend from the Atlantic coast to the
Mississippi River. In returning toward the north,
on the west side of the Alleghanies, we come upon
the forests along the divide between the head-
streams of the rivers that flow into the Gulf and
those that run into the Ohio, comprising portions of

Tennessee, Alabama, and Kentucky, and into West Virginia, so marvelously rich in her native timber; and across the river we find the eastern portion of the State of Ohio and Western Pennsylvania abounding in remnants of forests that find their northern limit on Lake Erie. Farther west, north of the prairies, and along and over the divide toward the lakes, and on the head-streams of the Father of Waters, are some of the finest forests within our land, and to the south in Eastern Missouri, in Arkansas, and Texas. The trees of the latter two partake of the general characteristics of those found in the other States in the same latitude. Texas has forests of pines whose trees are often one hundred feet high, while she takes pride in her pecans, as being indigenous to her soil. This is a species of hickory, "the wood coarse-grained, heavy, and durable," the trunks straight and from fifty to seventy feet high; the nuts are almost cone-shaped, are fine-flavored and of good size. The old trees in the forest bear abundantly. But Louisiana, also, claims this tree as indigenous to her soil and "the pecan as the richest and most delicate nut in the world." The tree in that State "attains an enormous size, often measuring fifteen feet in circumference, the height reaching to 125 feet, and the shadow at noonday covering a circle 115 feet in diameter." As the tree grows older, its production of nuts increases from year to year. (*Louisiana, Resources, etc., p. 102.*)

In this connection we notice that amid the Rocky

Mountains are scattered groups of forests, here and there, while New Mexico and Arizona lay claim to timber-lands, but of limited extent when compared with other Territories farther north.

Pacific Slope Trees.—On the Pacific slope are the most prolific forests in the Union, as seen amid the foot-hills of the Sierra Nevadas or their northern portion, the Cascades, or along the Coast Range and in the valleys intervening, and also on the slopes of Alaska that border on the Pacific. We include among these trees the famous mammoth or giant trees of California; the latter are in two prominent groves—Calaveras and Mariposa. These marvelously large trees are protected by law, to preserve them intact forever, as there are none such in the world. By far the most valuable tree of California is the redwood, which belongs to the same class as the mammoth, though of a different species. Groves of redwood are found on the western slope of the Sierra Nevada, and along the Coast Range. These trees are very large, and the wood is peculiar in its deep-red color. The wood of the main trunk is plain in texture, but the stumps and roots are much curled, like those of the black walnut, and are used by cabinet-makers in veneers for ornamental work. When polished and varnished it is nearly equal to rosewood in the richness of its tints, and in durability to mahogany, though it is much lighter and more easily handled. As these trees are in abundance and their wood adapted to the ordinary purposes for

which it is used, the lumber derived from these forests, if not wantonly destroyed, will be sufficient to
supply the needs of the people for generations. In
point of utility the immense trees so densely stud-
ding the soil of Oregon and Washington Territory
far surpass any others on the slope—the Oregon pine
being the most valuable tree in that region. Nor
must we omit to credit distant Alaska with extensive
forests of pines, cedars, cypress, spruce, fir, and hemlock, which are found along her southern and southeastern coasts. The latter strip of territory lies for
400 miles along the shore, and extends back on an
average 140 miles to the crest of the mountains, that
line dividing British Columbia from Alaska. Among
the trees of Alaska is a species of linden said to be
peculiarly well adapted for producing pulp from
which a fine paper is made.

Specimens collected.—In the city of New York,
the American Museum of Natural History has a
complete collection of specimens of the various trees
from the native forests of the United States. Here
are 417 specimens, each one so labeled that the visitor may ascertain its characteristics, and learn from
the map attached to it the section of the Union to
which it belongs. " These specimens are cut in such
manner as to display the bark, and cross and longi-
tudinal sections of the wood, one half polished and
the other in its natural condition." The Museum
owes this display of American woods to the public
spirit and benevolence of one of New York's no-
28

ble sons, Mr. Morris K. Jesup. Every intelligent American citizen, if opportunity serves, should, on his own behalf, visit this collection, that he may have a conception of the wonderful variety of our native trees, and the extent of territory that they occupy.

The Middle Belt of Trees.—In noticing these 417 specimens our limits will only permit us to speak, and that very concisely, of a few of the most valuable and therefore important, and also very briefly of the uses to which their woods are applied. There is a decided variety in the characteristics and the size of the trees indigenous to the United States. Through the middle section of the Union, extending from the Atlantic to the Pacific, and from the thirty-seventh to the forty-second parallel of latitude, the trees are, for the most part, not remarkably tall, but are large in their trunks and in their expanding branches, so that they stand comparatively thin upon the ground ; the compensation consists in the latter being, in consequence, the more easily brought under cultivation. We shall see that in both these respects there is a wide contrast in many of the trees, both north and south of the lines just mentioned. Throughout this middle section are the favorite homes, including the varieties of each, of the oak, walnut, hickory, ash, elm (the latter more in the eastern portion), spruce, hemlock, sugar-maple, the chestnut, etc. The two latter deserve a passing notice. The sugar-maple finds its greatest development, west and southwest of the Alleghanies, in the fertile valleys and in the rich soil of the hill-sides. Its trunks attain sometimes a diameter of three or four feet, with correspondingly large and expanding branches; this tree conferred immense benefits upon the early settlers and their descendants for nearly a century, as its abundant sap was their only source for sugar. But growing in the most fertile soil only, it has given way to cultivated fields, and the people now obtain their sugar from the sugar-cane of the South. In contrast with this immense tree, and its coarse and hard texture of wood, is the comparatively slender and fine-grained and compact wood of the rock-maple of Vermont and New Hampshire, where to-day its groves are preserved and flourish in the valleys and on side-hills and furnish a sugar delicious in flavor.

Chestnut; Hickory; Ash; Walnut.—The chestnut crosses the entire belt and extends south along the ridge of the Alleghanies to their southern termination, and frequently is found in scattered groups for scores and scores of miles, on the hills of their outspurs northeast, east and west, and southwest. This tree, upon the whole the tallest of its own region, furnished the early settler an easy-splitting and durable wood for the rails of his fence, while the locust afforded a more lasting one for posts, and was popular because of its grateful shade. The woods of this middle belt are uniformly hard, and are adapted to purposes that do not require long beams. The hickory, because of its strength and compactness, is used for spokes of wagon- and carriage-wheels, and for handles of the utensils belonging to the mechanic and the farmer; the ash, because of its strength and elasticity, is used for the long handles of agricultural implements, and in the wood-work of reaping- and mowing-machines. For fine cabinet or furniture work are used the cherry, the bird's-eye maple, and, when a rich brown color is desired, black walnut. In this connection it may be noted that large quantities of American walnut are exported to Europe, as it is of a clear and smooth grain, in texture, while we import the French and German walnut, because it is not straight in grain, but curled in texture. In both countries the plain walnut is decorated or trimmed with the fancy or curled according to the taste and skill of the artisan. The oak, the ash, the walnut, and sometimes the chestnut and the maple, are often utilized in the inner wood-work of churches and public halls, and also in that of private dwellings. These woods, being hard and susceptible of a fine finish, are used in their native color, while their beauty is greatly enhanced by receiving a polish. This enumeration contains only a tithe of the uses to which the woods of this middle section are applied.

The Northern Belt of Trees. — North of latitude 42° comes the pine, by far the most important family of our trees. This tree, being very tall, and comparatively slender, is peculiarly adapted for beams in large edifices or in ship-building and for masts of ships. It has small limbs, compared with the trunk, mere twigs; owing to this peculiarity of the branches, the trees can stand more thickly on the ground, and therefore produce to the acre more thousands of cubic feet of lumber, than any other tree. In a general view it may be remarked that the soil in which the pine flourishes is for the most part sterile, and is ill adapted to reward the labor of the husbandman.

In the northeastern portion of the Union, in the State of Maine, are very extensive forests of white pine, whose tall trunks loom up as straight as an arrow. Passing west on the same parallels, or nearly so, we come upon the forests of the Adirondacks with a sprinkling of similar pines; but farther west, in the States of Michigan, Wisconsin, and Minnesota, abound immense woods filled with pines. In the latter State, the fifteen counties in the northeastern portion bordering on Lake Superior, and in the eastern part, have one fourth of their area covered with forests of white pine. (*State of Minnesota, p. 8.*) We pass still farther west on the same lines, and we come upon the fertile soil and the noble forests of Oregon and Washington Territory. These both have the same class of trees, such as the pines, the sugar, the white and yellow—the latter two predominate; the red, black, and yellow fir; one species of the latter, known as the Lambert, frequently runs to the height of 300 feet; large cedars remarkable for the fine texture and fragrance of the wood; the redwood,

the hemlock, and the spruce. These names represent only a few of the prominent trees of these forests, in some portions almost as dense as those on the
banks of the Amazon. In some districts the trees
are covered with a moss of a peculiar orange-green
color, with which the branches are festooned, while
the ground underneath in certain seasons is frequently beautiful with blooming, modest flowers—
white, yellow, and purple. Lumber is sent from
Oregon and Washington Territory to the ship-yards
of Europe, to be used for spars and masts.

While the most valuable trees north of parallel
42° are the different varieties of the pine, yet the
trees whose habitat is below that line often creep
above it within the spaces intervening between the
three great pine-forests just mentioned. In a similar
manner the pines themselves are sometimes found,
though in a straggling manner, encroaching on the
middle belt.

The Southern Belt of Forests.—We have already
noticed the Alleghanies and their outspurs covered
with various classes of useful trees, penetrating toward the south along the western portion of the
Atlantic slope. We now find farther east a similar
belt of forest commencing on the confines of Virginia, and nearer the ocean and trending southwest,
parallel with the coast. This belt extends from parallel 37° across North Carolina, South Carolina, and
Georgia, then branching off in two directions—one
across Alabama into Mississippi, the other southeast

into Florida. This entire belt may be properly termed a pine region; since, with its number of minor varieties, that is its most important tree. This vast forest region—some 60,000 square miles in extent—is by no means limited to the pine, as there are also other varieties of trees; each class valuable in a degree, if not equal to the pine, and which fill their special office in supplying the wants of the people.

Varieties of Trees.—Along the entire coast are found in the lowlands the trees indigenous to such districts. Many of these are hard woods, as the white and the live oak—the latter so necessary in ship-building by furnishing the knees for the keels of the wooden ships; the palmetto, the persimmon, the spruce of three varieties, and two kinds of cypress, which next to the pine is deemed in South Carolina its most important tree: it is of rapid growth and attains great size, while it occupies only waste places and swamps; the two kinds of cedar: the red, found in the swamps of Florida, deserves further notice. It has the most delicate texture of any known wood, and is used specially for making pencils, and in this respect has virtually the monopoly of the world. When Faber, the celebrated pencil-manufacturer of Stein in Bavaria, had exhausted his stock of Florida cedar, after the commencement of the American civil war, he was induced to send agents wherever cedar grew, in order to obtain a supply, but found none to fully suit the purpose.

Farther inland flourish hickories of seven varieties; even the close-grained pig-nut of the middle belt grows very large, and loses but little of its compactness. The oaks, of more than a dozen kinds, grow on the high and steep ridges, and are aecompanied by the stately chestnut, the hemlock, the ash of four kinds, the elm of three, and the sycamore. In the fertile valleys of this higher region, especially in Alabama, flourish the sugar-maple, the wild cherry, and the black walnut—the latter having a texture equally fine with that of the middle belt. This is a mere glimpse of the numerous trees that are also found within the limits of the great pine region, but they are all needed and none are out of place.

The Southern Pine.—Below parallel 37° we meet with a pine somewhat different in its characteristics from that in the northern belt, inasmuch as to the latter's good qualities as lumber is added in the southern the valuable property of producing turpentine, tar, or pitch, and rosin. These several qualities make the pine the most valuable tree upon the South Atlantic slope. From the days of the early colonists to the present time the two Carolinas have derived benefits from the active trade in the products of these forests, such as lumber in different forms; tar, spirits of turpentine, rosin, and naval stores: the latter furnished these stores to England before the Revolution, and since then to both her and to the United States Government. " The value of the crop of naval stores produced in 1880 in North Carolina alone was about $8,000,000, while that of

South Carolina was estimated at one third the aggregate amount in the United States." (*Handbooks of North and South Carolina, pp. 329 and 606.*)

In this extended region the pine attains its highest perfection: it flourishes in sandy and for the most part barren soil; in some sections standing thickly upon the ground, in others more sparse; in the latter, owing to the smallness of its branches, the sun penetrates quite freely to the earth, which, being free from underbrush, a variety of native grasses abounds, upon which feed herds of cattle and flocks of sheep, as in some of the pine-forests of Alabama.

Of the ten varieties of pines—some of little importance—only two are pre-eminently valuable, the short-leaved or pitch-pine and the long-leaved or yellow or Georgia pine—the latter thus named because it was at first exported only from that State. From these two varieties, almost exclusively, are obtained the tar, turpentine, and rosin in the United States. The yellow or Georgia pine has in addition a special value because of its great hardness and compactness of texture, which makes it unrivaled for use in public buildings, as in floors, and in steps of stairways on which the wear is very great. " South Carolina has 20,000 square miles of pine-forests in the lower part of the State, which furnish the very best quality of yellow pine." (*Handbook of South Carolina, p. 605.*) North Carolina has 12,000 square miles of pine-forest of both kinds; Georgia, 15,000; Alabama, estimated at 12,000; and Florida, at more than 1,000, in addition to her famous cedars, and about two hundred varieties of other trees.

Tennessee and Kentucky Forests.—The State of Tennessee extends from the Alleghany Mountains to the Mississippi River. It occupies a plateau on the divide between the water-shed toward the Gulf and that toward the Ohio; along its north boundary lies the State of Kentucky. The two combined have more than 50,000 square miles of forests, abounding in noble trees. This area includes their plains and hill-sides and mountains; for the latter, though often rough, are for the most part covered to their very tops with a luxuriant growth of trees. Of these are numerous varieties, the oaks having about a dozen; the yellow and blue poplar; the yellow and white pine; black as well as white walnuts, maples, firs, ashes, etc., and others—making in all nearly a hundred varieties.

Apprehensions are often expressed lest the timber of the United States should ere long be exhausted. The salvation and increase of the forests or timber of the country depend upon those·who have the facilities to renew the forests by planting, etc. Meanwhile the wanton waste of the timber, so repugnant to common sense, must be stopped, and in time there will grow sufficient to supply for the most part the wants of the people.

Tan-Bark; Tannic Acid.—It is proper, in connection with timber, to mention another resource— that is, the tanning properties of the barks of certain trees, such as, in the north middle portion of the Union, of the hemlock and different varieties of oaks, and the chestnut-oak in the southern belt of trees. The strength and quantity of tannic acid in this class of trees are quite different. It is claimed that the tanning properties of the chestnut-oak which covers the coal-fields of Tennessee, Georgia, and Alabama, to the extent of more than 10,000 square miles, is one third stronger than that of the bark of the same kind of tree in the northern section. It is estimated that these groves of chestnut-oak would furnish annually 300,000 cords of tan-bark for an indefinite period.

The Cañaigre.—Tannic acid is also found in a plant-root—the cañaigre (canyaigre)—which grows wild in great abundance in portions of New Mexico, and which has been used from time immemorial by the Indians in converting hides into leather. The Department of Agriculture of the Territory obtained from a competent chemist an analysis of the root, which was found to contain about 24 per cent of tannic acid. The roots have the general shape of the sweet-potato, and are from four to eight inches long by about one inch diameter; they are of a dark-brown color, while the inside is of a bright lemon-yellow; in some respects the root resembles rhubarb and has the odor of madder. Each plant has from three to six pounds of root, and which, when it becomes dry, still retains its tanning properties, the latter not being affected by long keeping. At present this root can be collected in great abundance, and hopes are entertained that it will yet be domesticated and cultivated. (*New Mexico Illustrated, pp. 96, 109.*) American sumac gives of tannic acid from 24 to 26 per cent, while white oak gives about 9 per cent, and hemlock 9½ per cent.

XLI.

GRASSES.

WE are so accustomed to see the fields and meadows carpeted with beautiful green grasses, that we scarcely realize the vast benefits which the latter confer upon the people, by affording sustenance to millions and millions of domestic animals, upon whose existence and welfare the people themselves are dependent for so many of the comforts and even necessaries of life. These grasses were originally native or of spontaneous growth, and some have since been improved by cultivation.

In the territory of the United States, the portions covered by forests, wild or native grasses grow only in comparatively small openings, where the power of the sun's rays reach the earth; these isolated spots were termed by the early colonists natural meadows, and to-day, what are known as the prairies, though very extensive, are small when compared with the entire forest territory. The native grasses, unless domesticated, nearly always retreat before the farmer's plow, and their places are taken permanently by those that are cultivated.

Varieties of Grasses.—About a half-dozen in number would include all the varieties of grasses in the Union that are cultivated to much extent. One of the most useful of these, the *timothy*, does its share in the fields and meadows in pasturing stock, but is oftener raised only for hay; and it appears to retain the greater portion of its nutritious properties when thus cured for winter use. It was noted as a wild grass in colonial times, and was domesticated by Mr. Timothy Hanson, of Maryland, in which State, as far as known, it originated. It was called *timothy*, from Mr. Hanson's Christian name, though it has been sometimes improperly named herds-grass. Mr. Hanson took some of the seed to England in 1780, where it was introduced, and is still cultivated. When seeded in connection with red clover, the combination enhances the value of both the pasturage and the hay, but in the field it continues to flourish without seed-renewal much longer than the clover sown with it. Timothy can be seeded in combination with the cereals, such as wheat, rye, oats, etc., as well as in connection with red clover. No one of our grasses is so much diffused over the land, nor one upon the whole so useful, as timothy.

White and Red Clover.—When the forests are removed, oftentimes the white clover springs out of the soil spontaneously, as if the "seed was in the earth," and was only waiting for the rays of the sun to call it into life, and thus it is often seen in the sunny openings in the midst of the surrounding forests. There are, according to botany, fifty-nine species of clover, the most important of which is the *red* or trefoil or three-leaved. It is thus named from the color of its ball-shaped head or blossom. In this blossom, as well as in the smaller one of the white clover, exists a sweetness of taste similar to honey, and both flowers are favorites of the honey-bee. The red clover can be seeded down by itself, or in connection with the cereals; the second year it grows luxuriantly as pasture, or it may be cured as hay for winter use, and the cattle eat it as though they found it delicious.

Grasses amid the Pines.—The native grasses in some of the States on the South Atlantic slope come into use early in spring, as in the lowlands of the Carolinas and Georgia, such as " the piny woods-sedge," and " switch-cane," and even the " Spanish long moss all through the winter continues succulent and nourishing, and is eaten greedily by all stock." The wire-grass of the pine-forests is quite nutritious, though coarse and repulsive in appearance, and also the Bermuda-grass, which, in the opinion of some, is a worthy rival of the red-top, or of timothy. Bermuda-grass seems to be well adapted to sheep-raising. In this section lucern—" the queen of all forage plants "—is utilized as a companion of red clover. The former is high enough for cutting by the middle of February, and " has been proved by experiment to be the most nutritious of all green food for stock." (*Commonwealth of Georgia, pp. 348, 349.*)

The Blue-Grass.—This famous and valuable grass had its origin west of the Alleghany Mountains in the central portion of the State of Kentucky. The early settlers noticed in the forest openings a re-markable native grass that densely covered the ground. It was of a rich, dark-green color, with a bluish tinge—hence the name, blue-grass; and where it thus grew spontaneously is designated as the blue-grass region. This friend of the lowing herds wakes from its winter sleep in the first month of spring, and hastens to fulfill its office by covering the fields in green. "It spreads a verdure so soft and fine in texture, so entrancing in its freshness, that it looks like a deep-lying, thick-matted moss." This nutri-tious grass, adapted to all classes of stock—even the swine eat it with relish—covers all the stretches of woodland pastures, and over the meadows and lawns, along the lanes and in fence-corners, and in every place where it can get foothold.

The seed ripens in June; during July and August it appears to cease growing; it is very sensitive to moisture, and should rain fail, and the heat be great, it seems almost to die out, but, when the rains of closing summer or of early autumn come, it rapidly revives and speedily clothes the fields again, though with a slight diminution of the nutritious properties it had in the spring and summer. If the season is mild, it continues into early winter, and even for a time grows a little. The valuable properties of this grass are seen in their effects upon dairy products, and in fattening cattle and sheep. Enthusiastic turf-men affirm that "it gives solidity to bone and strength of tendon, firmness and elasticity of muscle, power of nerve and capacity of lung" to the race-horse.

The use of this valuable grass is rapidly extend-ing into all those sections of the country where the climate and the soil are adapted to produce its fruit-fulness. West Virginia and Missouri claim it as one of their indigenous grasses. "Blue-grass fairly lux-uriates in this deep, flexible soil, and is fast making conquest of forest and prairie. It has as fine growth here as in the far-famed Kentucky, and by right of con-quest is the successor of the native wild grasses." (*Handbook of Missouri, pp. 140, 238.*)

The Buffalo-Grass.—One American grass—the buffalo—appears to spring spontaneously from the soil, for a special and limited purpose, and when that is accomplished it disappears, as it is incapable of being domesticated like the timothy and the blue-grass. It is said to flourish best at an elevation of 3,000 feet or more above the ocean. Upon it the buffaloes live almost entirely—hence the name; the two are found together upon the Western plains, and as the one becomes extinct, the other disappears, but not before it has also, for some years, it may be, afforded sustenance to the domestic cattle of the set-tler in the vicinity. Buffalo-grass seeds in the root, and can not be transplanted, and as soon as the land of the prairie is broken by the farmer's plow, and it

is turned under, it dies, and the roots, matted together several inches thick, decompose and become
fertilizers. The line of this grass, extending north
and south for nearly 1,500 miles, is rapidly retreat-
ing before the clover, the blue-grass, and timothy;
the farmers can sustain many more domestic cattle,
on an equal area of land, than the original grass sustained of the buffalo, the deer, or antelope. This
native grass is from two to four inches high, but is
as thick, according to some authorities, "almost as
wool on a sheep's back." It has the remarkable
property of curing in the warm sun just as it stands,
and becomes virtually hay, but retaining its nutritious elements. On this hay the buffaloes and deer
feed during the winter, often scraping off the snow
with their hoofs. This grass is produced in great
abundance during the months of April, May, and
June; then come the hot suns of July and August,
which cure it, thus making a fine nutritious food for
the animals during the winter.

Other Wild or Native Grasses.—The immense
value to the American people of the wild native
grasses that cover the plains of the West may be
inferred from the multitudes of cattle which are
pastured upon them. Great herds, often numbering
many thousands, are raised in these regions, to supply the wants of the people in the Eastern States, or
to be sent across the ocean to the markets of Europe.
Nor is this all; for in districts that are suitable for
the purpose are raised on the same grasses many

millions of sheep to furnish wool and food for the people. In addition, these grasses, until superseded by those that are cultivated, sustain the stock of the settlers along a line running north and south nearly 1,000 miles. These plains, that furnish such an amount of food to this portion of our domestic animals, extend from the Mississippi River to the foothills of the Rocky Mountains, and from the Rio Grande to the British possessions — about 1,500 miles.

The native grasses of the prairies present numerous varieties; the number of these has been placed by Prof. Aughey, of Nebraska University, at more than a hundred and fifty: but that is the minute detail of the botanist. Though these differences exist, there is much similarity among them in respect to their intrinsic value as food for animals. For illustration, on the prairies of the four Northern Territories, we meet with what is called *bunch-grass*—thus named from the huddled manner in which it grows. In its composition are elements in common with the buffalo, but owing, perhaps, to the character of the soil in which it grows, it appears in a different form. One authority speaks of it as "combining the food qualities of both hay and grain, and which supports cattle, horses, and sheep the year round"; another characterizes bunch-grass as a "combination of buffalo-grass and oats. . . . It cures itself where it grows, and during the winter the snowfall in that region— the four Territories just mentioned—is usually light, so that domestic animals find feeding-ground in the severest weather." Montana, for instance, east of the advanced spurs of the Rocky Mountains, "is a vast pasture-land that formerly supported millions of buffaloes and deer, but is now being rapidly occupied for cattle- and sheep-ranges," so that stock-raising on these enormous natural pastures is becoming more and more important. On the fertile prairies and in the rich valleys of Dakota, "the grass grows luxuriantly, covering even the high summits and table-lands five hundred feet above the valley."

Farther south, we find in Utah extensive stock-ranges, though the native grass is not deemed quite equal to the buffalo and gramma of the plains east of the Rocky Mountains. The latter grows, as elsewhere, in bunches, and in apparently barren places: early in the season it cures standing, and still retaining its nutritive qualities. Colorado and Nevada, also, have a share of the same kind of wild grasses, but not to so great an extent. New Mexico has many thousand square miles that are covered with wild grasses, the gramma and its varieties, white and black; it belongs to the family of the buffalo-grass of the other Territories. It grows in bunches and is very nutritious, and furnishes food at all seasons for all classes of stock. "It is flowerless and seedless, and covers the broad plains and the mountain-sides with withered-looking bunches that seem to combine the qualities of grain and the best of hay in the greatest perfection."

Arizona has 60,000 square miles or more of excellent grazing-lands, out of her 114,000 of territory. Except in a few strips of the country and the strictly mountainous regions, there is no portion of the Territory without a growth of grass; though a very different impression has hitherto gone abroad. These grasses partake of the general characteristics of those found in New Mexico, both as to their appearance and remarkable nutritive properties, cattle thriving on it as much in winter as in summer. (*Handbooks of New Mexico and Arizona.*)

Alfalfa.—This is also known as Chilian clover, as it appears to have been introduced from that country; it is a very valuable grass, belonging to the clover family, and appears to be well adapted to Southern New Mexico and Southern California, though it seems not to have been introduced into Arizona. It is cultivated and yields in great abundance, and the stock not only eat it with avidity, but on it thrive and fatten. This clover-grass attains a height of from twenty to thirty inches, and during the season can be cut several times, and is of course the most available green forage in the summer. "Alfalfa, when in blossom, from May to September, affords the best of pasturage, not only for stock and swine, but for the honey-bee." (*Handbook of New Mexico, p. 109.*)

In Southern California this kind of clover is much cultivated, and, with proper irrigation, produces three crops a year, and averages from six to eight tons to the acre, while the white and red clover will yield but one crop a year. The alfalfa "will produce more than double that of any other clover known ; and, if the quality is not quite equal to the red and the white, it certainly makes a very good feed in far greater abundance." It is fed to work-horses and milch-cows ; and swine, even, keep fat on it in that climate.

XLII

ORCHARD-FRUITS.

THE numerous fruits, great and small, of the United States add immensely to the means of sustenance and comfort of the people; these, taken as a whole, include the fruits of the temperate zone, and also those of the sub-tropical. Owing to our increasing and rapid inter-communication between the different portions of the Union, by means of ocean-steamers, of steamboats on the rivers, and railways, the products of the fruit areas of the country are made available for all the people. Among these, that prince of orchard-fruits, the apple, because of its many excellent qualities, stands pre-eminent in value. It is the most abundant in production and the most useful, and confers more benefits upon the people than, perhaps, all the others combined. It contains elements that are conducive to health and nourishment, and is susceptible of being prepared as food in numerous ways. The comparatively long life of this tree, and the almost endless varieties of the fruit —amounting to several hundreds—greatly enhance its value. Certain kinds ripen and become mellow on the tree during the summer and the autumn, while still larger and varied classes, which in the end are the most useful, after having had their perfect growth, and having ripened on the tree, can be stored away to develop their juices, not meanwhile deteriorating, but improving in flavor, and becoming luscious to the taste, while retaining their peculiar fragrance to the end. The process recently (1876) introduced, of evaporating in a few minutes the moisture of the apple without depriving it of flavor, greatly increases its value. The less choice apples are made into cider for domestic use, or to be converted into pure vinegar. The property of being kept in store enables the merchant to transport apples to long distances in our own country, and even across the ocean. This is in marked contrast with our sub-tropical fruits, the bananas, oranges, etc., which are so perishable by nature that they must be taken from the tree and sent to distant markets before they are fully ripe, and have perfectly developed their juices and flavor; in cousequence, after their long journey, they are found to be somewhat deficient in these desirable qualities.

The Apple Belt.—Consistent with the great usefulness of the apple, is its wide diffusion over the Union. The apple belt extends across the continent, from the Atlantic to the Pacific, and in the northern and middle portion without a break, while it stretches far south on the highlands and plateaus to such an extent as to be virtually universal. Though the apple develops best in the northern and middle portion of the Union, yet a few of the summer species are sometimes found in the vicinity of the sub-tropical fruits, as on the lower portions of the South Atlantic slope. There is not a single State or Territory that is devoid of the apple.

Johnny Appleseed.—Gratitude forbids us to forget the humble services of Johnny Appleseed. This name, so characteristic, was given him by the early settlers west of the Alleghany Mountains, because of his custom of planting the seeds in open spaces in the wilderness, that apple-trees might greet the settlers when they came. His real name was Jonathan Chapman. He was born in Boston, in 1775, and in early manhood appeared in Western Pennsylvania. He obtained his seeds from the cider-mills in the older settlements of the latter State, and then, passing over the Ohio River into the almost unbroken wilderness, he often carried these seeds himself, though sometimes using a horse. When he wished to reach settlements on the banks of the river farther down or west, he would load a canoe with seeds, and float it down to them. No doubt the efforts of this eccentric but kind-hearted man resulted, directly or indirectly, in great benefits to the three States that now occupy the peninsula between the Ohio and Mississippi Rivers. Says the "Grange Visitor": "Johnny Appleseed laid the foundation for the immense growth of fruit-trees, whose yield to-day forms so important a part of the annual products of the great State of Ohio." He began his work in that State in 1801. He scorned to be dependent, but supported himself. To accomplish this, he would sometimes select a suitable opening in the woods, where the sun's rays reached the earth; there he would plant his seeds, and, to protect the young trees, inclose the place with a brush fence. In due time he would return and sell these young trees to the settlers, in order to be independent, and also have the means to prosecute his benevolent work, and thus he labored during very many years. If a settler were really too poor to purchase, he presented him a number of trees. Numerous places that he thus planted he never visited again, but left them for the incoming settler.

The Peach.—Next in importance to the apple comes the peach, a luscious and delicate fruit and of a peculiar flavor. It originally came from Persia. The blossoms are very tender, and liable to be injured by frosts late in the spring, or by sudden changes of temperature, and the buds by very cold weather even in the latter part of winter. The trees are frequently a prey to numerous insects, and upon the whole are also short-lived. These conditions render the peach-crop, perhaps, the most uncertain of our orchard-fruits. In its native state the peach soon decays, and, in order to keep it for future use, it must be preserved by canning, in which process its flavor is somewhat diminished; while it is used only as a relish or delicacy. The new process of

rapidly evaporating the moisture of the peach, retains its flavor, and thereby greatly increases its usefulness, as in the case of the apple.

The peach-tree, under some mysterious influence,
is liable to disappear from whole sections of the
country, or linger on in a half-living and unproductive condition; and yet, after the lapse of a third or
a half century, a different variety, when introduced,
will sometimes flourish almost equally with the original stock.

The Peach Belt.—The peach area on the Atlantic slope includes, to-day, principally portions of the
States of New Jersey, Delaware, Maryland, and
Pennsylvania, though within recent years the fruit
appears to be gradually occupying districts where it
flourished half a century ago, as in certain portions
of New York State. Farther south, along the eastern foot-hills of the Alleghanies, the peach grows
and produces sufficient to supply domestic wants,
and in portions of the Carolinas, and on the highlands of Alabama, Tennessee, and Kentucky, but in
limited quantities. The State of Georgia cultivates
a very large area in peach-orchards, and produces a
fruit of a remarkably fine quality. The Horticult-
ural Society of the State reports that within its
limits are fifty-five varieties of the peach. " The
advantages possessed by the peach-grower of Georgia far exceed those enjoyed by the orchardist of any
other State in the Union." (*Commonwealth of Georgia, p. 336.*) The peach ripens early in that section

and becomes fully developed, but the value of the fruit to the people at large can be greatly enhanced by the introduction, where the peach grows and ripens according to nature, of extensive canning and evaporating establishments, by means of which the delicate flavor of the fruit can be retained. In this State are peach-orchards, great and small, reaching as high as 70,000, in their number of trees. Florida is also engaged in the industry of peach-growing. Says the Florida Times-Union: " The wide-spread interest in the peach and its productiveness promise to be as remunerative in yield as the orange or the lemon."

The Peach in the West.—In the States and Territories within the valley of the Mississippi, except in the northern part, the peach is almost everywhere found in company with the apple, though its relative production is not equal to that fruit. Texas in the West, like Georgia in the East, is the first in the season to produce the peach in perfection. Says the Galveston "News": "By the first of June the people of North Texas are enjoying the luxury of that luscious fruit, the peach."

The State of Missouri is also specially productive in peaches. The cultivation of this fruit has increased in California at a manifold rate, and that in districts once thought to be unfavorable to its production. The fruit is of an excellent quality, as, for instance, that grown on the foot-hills of the famed Santa Clara Valley. These peach-orchards supply the Rocky Mountain region and the northern portion of the Pacific slope.

Pears, Plums, Cherries, and Quinces.—These four fruits are so well known and so extensively cultivated, that to go into details concerning them would seem superfluous. They are specially valuable, as they are so capable of being preserved in cans and in such manner as to retain much of their original flavor. The pear is usually considered second in value to the peach : its varieties are numerous, and they are found in different sections of the land, as the tree has the unusual power of accommodating itself to its surroundings ; thus, different from the apple and the peach, it flourishes in the sea-coast country of the South Atlantic, and also on the highlands in the interior. The pear is propagated by cuttings or grafts, and the skillful fruit-grower can thus reproduce to any extent the peculiarities of the specimen selected. For illustration, it is recorded of a tree—the first bearer of the pear known as Le Conte —in Georgia, "from which hundreds of thousands of trees have been propagated." (*Commonwealth of Georgia, p. 337.*) The three foremost States in the production of pears are California, Georgia, and Florida. The former also takes the lead in apricots, cousins of the plum, and also nectarines, bearing a similar relation to the peach. The California pear is deemed by many the best fruit grown in the State ; these pears in vast quantities are consumed in the cities of the valley and in those east of the Alleghanies.

Plums flourish throughout the temperate zone, wherever the apple or the peach grows. In some sections, apparently from local influences of climate or soil, they are more productive than in others. The varieties of the plum are numerous, and skillful nurserymen are steadily increasing their number and also improving their quality. The *cherry* is also widely diffused, though its importance as food is not equally great with either of the two former. Its varieties are not more than half as many as those of the pear, but its value is much enhanced by its capability of being easily canned or preserved for future use. The *quince* stands pre-eminent as a fruit susceptible of being prepared and used as a preserve ; and, as the tree is less productive than any other, the peculiarly rich flavor and pleasant taste of the fruit compensate for the deficiency.

THE ORANGE.

The orange is not indigenous to the United States. The trees producing the Seville or sour orange were brought to Florida by Spanish Catholic priests or missionaries, about the year 1570, and the tree and its fruit were scattered over the region by the Indians. The sweet orange was unknown in Europe at that time, but was introduced there, about thirty years later, probably from Asia Minor or the Grecian isles. The sweet-orange trees were afterward, also, brought to Florida by Spanish colonists, and planted in the vicinity of St. Augustine ; thence they were carried to the settlements on the St. John's and Indian Rivers. The pollen and the flowers of these two species—the sour and the sweet—in the course of time mingled and produced that hybrid, the bitter-sweet orange; though the former two prevail in the forests, and still preserve their individual characteristics.

In the lower counties of Georgia are native groves of sour-orange trees. In the same localities sweet oranges of fine flavor and size are easily raised for domestic use, and also on the sea-islands on the coast, though the sweet orange finds its better development farther south, in Florida.

Localities of the Orange.—Three sections of the United States are prominent in the production of the orange: a portion of Florida, of Louisiana—the trees of the latter derived, evidently, from the former —and of Southern California. The area in Florida, possessing the best conditions under which the orange can be cultivated to perfection, is estimated by a careful authority at 10,000 square miles. The sweet orange raised in this State is remarkable for its excellent flavor, so that in size, in production, and general good qualities, it is pre-eminent. " Taken from the tree in full maturity, the Florida oranges surpass in excellence all others wherever grown. . . . Their unusual keeping qualities enables the grower to continue the fruit season six months." (*Florida Times-Union.*) On these trees may be seen at the same time the blooming blossom, the green fruit and the ripe. The groves are beautiful beyond compare; the golden fruit and the white flowers appearing amid the green foliage, while a delicious fragrance penetrates the whole atmosphere. Each tree during the season on an average yields two thousand oranges, and one acre, when properly managed, has been known to produce ten tons of oranges. The growing of this fruit is a very important industry of the State; the modes of cultivation are constantly improving, and also the methods of propagating the best varieties. The tree itself will live about one hundred years, and continue bearing fruit to the end. The first varieties are often propagated by seedlings or by budding the choice stock on the long-lived trees of the wild grove. The scientific cultivation of the orange is in its youth, if not in its infancy, and experiments by enterprising growers are gradually improving the qualities of the fruit.

Louisiana Oranges.—These partake of the general qualities of those of Florida; however, the yield is enormous, said to be much larger to the tree than in Florida, but whether the flavor is equal is undecided, as the orange, because of its delicate constitution, is easily affected by climatic influences. The favorite locality of the orange in this State is a district commencing on the west bank of the Mississippi, about forty miles below New Orleans, and thence extending to within a short distance of Fort Jackson. There is along this bank for thirty miles an almost continuous grove of orange-trees, the number of which in separate·orchards ranges from a few score up to ten thousand. As an evidence of this being a genial home for the orange, there are, it is claimed, trees in this district more than a hundred years old, and which are still producing fruit.

As Florida sends supplies of her luscious fruit to the cities and villages on the Atlantic slope, so Louisiana performs a similar service for the people dwelling in the valley of the Mississippi; and California, also, sends thither car-loads of oranges and other fruits to aid in the good work of supplying the great cities and villages of the upper portion of the valley.

California Oranges.—The counties in the southern part of this State produce immense quantities of oranges and other sub-tropical fruits. These fruits are marvelously large, but said to be deficient in flavor when compared with those raised in the East, though the latter are smaller in size. The soil and the climate of this region appear wonderfully adapted for the production of fruits of various kinds, not only in large quantities, but in unusual perfection. In cultivating the orange, irrigation is often resorted to with great advantage. The first portion of the orange-crop is usually ready for market from the 1st to the 15th of December. Recent improvements in the internal arrangements of refrigerator - cars enable the merchant to preserve fruits, otherwise perishable, so that they can be carried great distances in as perfect condition as when taken from the

30

tree, the car remaining unopened till it arrives at its destination. The cultivation of the orange has within recent years been gradually extending into the middle and north-middle of the State in the valley of the Sacramento River, where the orange-tree is found flourishing in hundreds of places, while the fruit produced is claimed to be equal to that of the southern portion of the State. The amount of production will, no doubt, continue to increase in proportion to the demand, until the advantages of having fresh and wholesome fruits can be enjoyed by all the people, instead of being limited so much to the locality where they are grown.

Minor Sub-Tropical Fruits.—To these belong the lemon, citron, and the lime, of the same general class; the fig, the cocoanut, the pineapple, and the banana. The lemon somewhat resembles in color the orange, but is oval in shape—its pulp is intensely acid; the citron is nearly the same in character; the lime is smaller, and round in shape and more acid than the lemon. On the Atlantic slope, Florida is the leading State in the production of sub-tropical fruits. Into the southern portion have been introduced, from the Bahamas and from Cuba, the banana (though it is not equal in flavor to the same fruit of the tropics), the pineapple, and the cocoanut. Improvements in the cultivation of sub-tropical fruits are now increasing in this State from year to year, the industry not receiving hitherto the attention that its importance demanded, though Nature supplied a soil and a climate as a natural resource of great value to the people if properly utilized. " The cultivation of cocoanuts and pineapples, is becoming an important industry upon the Keys adjacent the Florida reefs." The former were introduced in 1845, and the latter in 1867 by islanders from the Bahamas, and their culture is gradually extending northward in the State.

" Though of such easy and sure culture, no available method of curing the fig has been introduced." Immense crops can be raised in Southern Georgia and in Florida, the quality being as good as that imported from Smyrna, but the producers have yet to learn how to prepare them for market. (*Commonwealth of Georgia, p. 341.*) The lemon and the citron are found in Louisiana, in company with the orange, and also in California. In the latter State is cultivated the English walnut, and the olive-orchards are large and bear fruit of fine quality—"plump, juicy, and full of flavor," equal to those of France or Spain —and the oil produced is of the best quality and unadulterated ; and here are, also, cultivated hard and soft shell almonds.

XLIII.

THE grape is indigenous to the United States, and, when the English colonists came, the vine flourished in the forests—not so vigorously in the northern portion of the present territory of the Union as in the southern. In certain portions of the latter region the trees were festooned with vines loaded with rich clusters of grapes. Some of these were of indifferent quality, and others were rich and juicy; the latter in time were domesticated by the colonists, and have been a source of food from that day to this. A remarkable one among these was the " Catawba, of a light claret-color when ripe "; used for making wine, as well as a table-fruit. This grape had its home in the southern portion of the Atlantic slope. In the same region, but more on the higher lands, was the scuppernong, a large round grape, growing wild from Virginia to Florida; a valuable species, similar to a native grape of Greece, and now extensively cultivated. West of the Alleghanies, amid others that were much inferior, was the fox-grape, in form and color similar to the scuppernong, and was perhaps of the same species.

The domestic grape is now so universally cultivated in every portion of the Union, that there is not a State nor a Territory in which it does not more or less flourish, though the climate and soil of some districts are more congenial than others to the growth and perfection of its delicious juices. The markets in our cities testify to the fact, by their exhibitions of grapes, that within the last half-century the grape has been improved by culture more, perhaps, than any other of our fruits, not only in the enormous increase of production, but in a higher grade of excellence.

Great changes have taken place in this form of industry, and sections of the country, in which it was once thought the grape could not be cultivated to advantage, now produce immense quantities. For illustration, we may refer to the valley of the Hudson, or the lake-region of the State of New York, or to the State of Ohio, taking the vicinity of Cincinnati as a central position, from which the cultivation of the grape not only extended within that, but in the neighboring States. In truth, our entire industry of grape-culture, though it has thus far been successfully prosecuted, is in its youth, as, in addition to the vineyards, in which the raising of the fruit is carried on as a business, the cultivation of the grape has spread from farm-house to farm-house all over the land.

The Special Grape Belt.—Upon the whole, it would appear that the most congenial climate and soil in the Union, for the cultivation of the perfect grape, is that belt of territory including New Mexico, Arizona, and Southern and Middle California. The former two are only partially developed, but, as an earnest of the future, they manifest in every respect an unusual adaptation for grape-culture. " The Mesilla Valley, in Southern New Mexico, is said to produce a grape with juices heavier than those from the grapes of Madeira and Portugal, as the grapes remain on the vine until they commence to dry before being pressed; and the wort contains as much sugar as the sweetest of Malaga. When dried they make a good raisin." (*Dr. Dennison, Health Resorts, p. 15.*)

In this vast region the grape was introduced from Europe two hundred years ago, by Spanish Catholic priests or missionaries, and the immense . growth of the vines themselves, as seen to-day, and their vast production of fruit, bear evidence to the adaptation of the soil and the climate to produce the perfect grape. As an evidence of what may be expected from this climate and soil, there is, near Santa Barbara, a vine which was planted by a Spanish priest half a century ago. The stem of this vine is more than three feet in circumference; it is properly trained on trellis-work—covering an area of nearly 5,000 square feet—and has been known to produce in one year more than five tons of grapes, some of the clusters weighing from four to five pounds!

California has made rapid and sure advances in this industry; she laid under contribution all the prominent vineyards of Europe in order to secure their best varieties, and even has partially repaid the benefit by sending back cuttings of the same, but which were much improved by a sojourn in her genial climate. There is no fruit so sensitive as the grape in respect to its surroundings, be it of climate, of soil, or of location, and no one that requires so much care in securing the proper conditions. The latter being known, the way is open for the grape of the United States to improve from year to year in good qualities, and in an enlarged production, that will supply the demand of an ever-increasing population.

Locations of Vineyards.—In the northern portion of the Atlantic slope, New York takes precedence in having vineyards; in the southern, Georgia. The latter has thirty-three varieties of the grape, and it is the only State in that region that has vineyards to any extent. "From her mountains to the sea-coast line the scuppernong is brought to perfection. It makes several distinct types of wine, all highly perfumed and of delicious bouquet, and a brandy of unequaled excellence." (*Commonwealth of Georgia, p. 339.*) In the Great Valley, the State of Ohio has precedence in the number of her vineyards; how long she is to maintain that pre-eminence remains to be seen.

California exceeds any of the other States in the number of her vineyards and in general grape-culture. Experience has taught her grape-culturists many useful lessons that will enable them to succeed better in the future. Experiments are being made that will no doubt result in still greater progress. This State has more than 250 varieties of grapes; these are the outgrowth of the numerous cuttings or specimens of vines that have been brought from every grape-growing region in Europe. When these were placed under different influences, of both climate and soil, the sensitive grapes became modified, often in their color, as well as in their juices and general characteristics. This great number of varieties would seem improbable were it not established by fact, but only those that have been tested and proved to be choice are cultivated. Those only that are specially fitted for table-use find their way in refrigerator-cars to the distant cities, and thus millions can enjoy the benefit of this delicious and wholesome fruit.

Raisins.—Large quantities of the best grapes are prepared for market in the form of raisins. The process of raisin-making demands extreme care, and a thorough knowledge of it is difficult to attain; and the impulsive Californian was long in acquiring the requisite patience and experience. There are several conditions of the grape that must be considered—the period when its juices are in the right state to develop into the perfect raisin, neither too dry nor too moist, but just right—in order to preserve its original flavor. This industry at present consumes but a small proportion of the State's grape-crop. In 1885, 400,000 boxes of raisins found their way to consumers outside; but the indications are that the number of boxes thus prepared will speedily reach many millions, as the industry is scarcely introduced to any large extent.

Wines.—The much greater amount of grapes raised in California is used in wine-making. To produce wines similar to those made in Europe, seedlings or cuttings of vines producing the same kinds of grapes are imported, and in the same manner for the manufacture of brandy. It is claimed that California has extensive districts wherein grapes are cultivated which contain within their limits similar, if not more genial climatic characteristics than obtain in the grape-producing countries of Europe, and also qualities of soil similar to those found in these localities. (*San Francisco Chronicle, January 1, 1886.*)

It is also claimed that the wines thus made are absolutely pure; if for no higher motive than that it is cheaper, because of the vast abundance of the grapes, to make the wines pure than to adulterate them. A highly respectable French gentleman, a wine-merchant in the city of New York, said to the author, in speaking of the wines of California: "They are in the main as good as those of France, and of one thing the buyer may be certain—they are pure." Thus far (1886), at least, this is their reputation.

Small Fruits.—These include the garden productions, such as peas, gooseberries, currants, etc., and berries of every description, many of which grow spontaneously in the woods and native meadows, and some on comparatively barren soils, as the whortle or huckleberry and the blueberry, while the cranberry flourishes in marshy districts, where it has been partially cultivated; but the most important of all, the strawberry, that prince among berries, when in a wild state, is found only on fertile soils amid the luxuriant grass. If we were deprived of these small fruits for a season, on their being restored, their worth would be appreciated more than ever. The value of these fruits is in their imparting a wholesome stimulant to the system, an enjoyment, as well as a promoter of health.

Berries for Birds of Passage.—Many species of these wild berries indirectly confer great benefits upon the people in ways not generally known. For illustration, there are almost unlimited quantities of a species of cranberry and of strawberry that flourish in their season in Alaska, and their vines cover many hundreds of square miles on the shores and marshes adjacent to the Yukon River. There in the summer-time, for three or more months, these berries alone furnish food to immense multitudes of wild ducks—the famed canvas-back and others—and wild geese that, guided by instinct, come there to the far north, from their southern places of resort on the bays and inlets along the Atlantic and the Gulf, from Delaware Bay, round to the mouth of the Rio

Grande. They come thither to lay their eggs in
nests amid the sedges that extend for hundreds of
miles along the shores of that river, and to hatch and
rear their young. When the latter are grown sufficiently large to fly, the parent birds lead them to
the bays and inlets just mentioned, to become a
prey to the wiles of the sportsman and the hunter, and to furnish during the winter the markets
with delicacies for the table. In addition to the
numerous wild berries found on the Yukon River,
and which appear to have been provided for a
special purpose, it is worthy of notice that the
more southern portion of this Territory also produces spontaneously, in their season, delicious wild
berries of many kinds and in very great profusion.

Northern **Wild Berries.**—In the northern portion
of the Union the wild small fruits are indigenous to
the soil, and grow very vigorously in the thin woods
and in their open spaces. The wild blackberry and
the raspberry—white, red, and black—are found in
great abundance in the middle and north middle part
of the United States, extending in a belt entirely
across the continent. They afford a large amount of
table relish and food for the people, and sustenance
for the winged game of the woods, and thus indirectly benefit the people, as this game is often secured for food. The small wild fruits of the southern portion are not so vigorous in growth. The
wild blackberry, the raspberry, and the strawberry

have been domesticated, and thus their usefulness has been increased many fold.

The Lawton Blackberry. — A gentleman near New Rochelle, New York, noticed, in a secluded part of his farm, a single blackberry-bush, having a large berry, which had a peculiarly pleasant taste and fine flavor. He transplanted the bush to his garden, where it flourished finely and produced an excellent fruit and in abundance. From this plant was propagated the Lawton blackberry — thus named from the gentleman who recognized its good properties and made it available for popular use. A similar instance occurred in Pennsylvania, in which an extra-fine specimen of the black raspberry, indigenous to that State, was thus utilized and propagated. The wild red raspberry prevails more than any other in the northern portion of the country.

The Strawberry. —No one of these, however, compares with the strawberry in its aromatic flavor nor in its pleasant and mild acid taste. This plant welcomes the sun and blooms at the opening of spring in Middle Florida, where the fruit ripens in April; but, still responding to the warmth of the sun as he moves northward, it continues to bloom and bear till toward the first of August, when its berries disappear from the woods and gardens of Canada. This plant, having been for a long time domesticated, has afforded in its season the best small fruit known to American tables. The varieties of the strawberry are numerous and increasing in number, and their good qualities are retained, as, after being tested, the best are selected for cultivation. Some of these grades may be slightly more fragrant and luscious than others, but the latter often compensate by being hardier and in being more prolific. Fast railway-trains bring strawberries from Florida in April to the main markets in the cities on the Atlantic slope, while toward the last of July they are brought from Northern New York and Canada. The time was when one month covered the fruit season in the city of New York; now the railways extend it to nearly three months. This may be said of all these fruits, wild or cultivated. Meanwhile the refrigerator-car from distant California distributes in their season the admirable fruits of that State in the cities of the Great Valley, and often in those east of the Alleghanies.

The Melon.—Melons of both kinds migrated during the ages from India to the Levant; thence to Europe, and thence to America. There are two general classes—the water- and the musk-melon ; the latter thus named because its fragrance is similar to that of musk. Of these two classes are numerous varieties, all of which, under the hands of the skillful gardener, are susceptible of future improvement in size and flavor. This delicious and refreshing fruit in both kinds is cultivated throughout the Union from ocean to ocean, wherever the climate permits ; the latter fails only in the extreme northern portion to cherish their growth. On the Atlantic slope,

Georgia is the most prominent State in producing the water-melon, while New Jersey—the Garden State—cultivates both classes, but is deemed the first in respect to the musk-melon. Other States, as Maryland, Virginia, and Delaware, also aid in furnishing the markets along the Atlantic coast.

The Results of Rapid Trains.—Somewhat inland, but all along the Atlantic coast from Florida to New York, abound gardens that produce early vegetables and small fruits. The remarkably genial climate and fertile soil in the more southern end of this belt aid much in supplying these wants of the people farther north. Norfolk, in Virginia, owing to its central location near the coast, and also to having outlets, both by steamship and railway, has become the main depot for the small fruits and early vegetables grown in Virginia and Maryland, and also for those raised farther south, in the Carolinas and Georgia, which are thence carried for distribution in the Northern cities. These advantages of climate and soil, thus connected with rapid communication, make this resource of immense value and of ever-increasing importance to the people. The upper portions of the Mississippi Valley can be more or less supplied with small fruits and early productions of the garden from Louisiana and California.

Fruits can be improved.—What a wonderful economy the All-wise Creator has instituted, in which fruits of nearly every kind can be improved by proper cultivation! This (shall we not say divine?) law of Nature opens a wide field for success in cultivated fruits. The United States, in consequence of their extensive territory and varied climate, are peculiarly blessed in their fruits, great and small; while, still more, the improvement by culture of these fruits has a bright future.

The seeds of orchard-fruits never produce one precisely like the original from which the seed is taken, but instead a different variety. The latter may be inferior in quality to the original, or it may be superior, but neither is known till the fruit of the tree produced from the seed is tested; and if it is inferior, the tree is thrown aside as useless; but if worthy of preserving, its variety is propagated by cuttings or grafts. This process requires time and care, and the judicious nurseryman is always on the lookout for these superior specimens, be they of apples, peaches, plums, pears, apricots, oranges, etc. The same principle prevails in improving the different varieties of the small fruits and all classes of berries. The latter are propagated by roots or fibers, not by the seeds. For instance, the *strawberry*-seed is properly planted and carefully watched. The new plant is transplanted, and the fruit it bears noticed: if it is inferior, the plant is destroyed; and if superior, it is preserved carefully and propagated by the roots or fibers. The variety of fruit thus produced retains the same flavor and characteristics, and, if excellent, obtains a wide diffusion.

XLIV.

IT is consistent with the goodness of the Creator to provide a variety of means by which to enhance the fruitfulness of the earth, upon which man chiefly depends for sustenance. For illustration, when its powers of production have been impaired, they can be renewed through the medium of certain substances that may have been in store for long ages. Fertilizers perform two offices: one, to supply the soil with needed elements; the other, to render available certain constituents that are already in it, in order that they may be assimilated as plant-food by the fibers or roots. Chemistry has analyzed the several grains, and revealed the ingredients of which they are each composed; and the soil that is deficient in an element which is essential to the perfection of any one class of grain will produce a defective *kernel* of that class. The intelligent and praetical farmer or planter endeavors to supply the deficiencies of the soil he cultivates with fertilizers containing the appropriate elements, in order to produce the *perfect kernel* of the grain he wishes to harvest, the finest fiber of his cotton, or a delicate texture of his tobacco.

Marl.—Marl is a deposit of sedimentary and mixed earthy substance, consisting usually of the carbonate or phosphate of lime, mingled often with clay and sand, and with vegetable and animal matter. The former two, originally held in solution by the water when in motion, were precipitated when the latter was still, as in shallow lakes or marshes. A belt of territory, with an average width of from twenty to thirty miles, containing a series of large but not closely connected beds of marl, commences not far inland from the ocean, in Southeastern Virginia, and extends in a southwesterly direction and parallel with the coast, across the States of North and South Carolina into Georgia, where in the latter State, leaving the coast, but keeping the same direction and retaining nearly the same width, this belt of marls continues across the State of Georgia, reaching the Chattahoochee River some distance below Columbus; and thence, passing through portions of twenty counties in Alabama, it crosses the State line into Mississippi, directly west of Grove Hill in the former State. This is the general outline of these singular deposits.

Animal Remains.—These separate beds of marl along the coast contain animal remains, such as of sea-shells, the bones of fishes and of sharks, estimated to have been from forty to sixty feet in length, and partially aquatic creatures, such as the tapir, the seal,

31

the walrus, and dugong, or sea-cow, and also of land-animals, such as the deer, the mastodon, or elephant. The substance of the decomposed bodies of these animals had influence in determining the character of the marl. These classes of remains are found especially in the marls of the Carolinas, where the water was no doubt saltish and attracted land-animals. In Georgia and Alabama the "shell-marl" predominates, there being in it no remains of land-animals, but "the shells or secretions of marine creatures." Marls have often characteristics slightly different, and these are designated by appropriate names—as greensand, shell-marl, and blue-marl; in South Carolina are two extensive groups named from the rivers that run across them—the Santee and the Ashley-Cooper. It is remarkable that, in sinking artesian wells in the South Carolina districts, marls and animal remains have been found 700 feet below the surface.

The Extent of Marl Area.—North Carolina has about 12,000 square miles of marl territory—that is, where it is found; this includes portions of twenty-five counties in the eastern part of the State that lie along the coast and extend from thirty to sixty miles inland. The marl in this State lies in horizontal layers, and frequently crops out on the sides of ravines or gullies and river-banks, in connection with an impure limestone, from two to four feet in thickness. The most efficient fertilizing agent in these marls is the lime they contain, which often varies in amount, running from 10 up to 95 per cent.

South Carolina has also a large area of marl, over a portion of which lies the phosphate-rock—to be noticed presently. In Georgia a belt of isolated "shell-marl" beds extends from the Savannah River, commencing a short distance below Augusta, in a southwest direction across the entire State and through portions of eighteen counties, to the Chatta-hoochee, and, crossing that river below the city of Columbus, the belt continues across Alabama, and over the State line into Mississippi, directly west of the village of Grove Hill in the former State. The area of marl in Georgia and Alabama combined is about 1,200 square miles, of which 550 belong to the latter, a portion of which has also phosphatic rocks. (*Georgia Commonwealth, Handbooks of North and South Carolina, and Description of Alabama, article Marl.*)

Phosphate-Rock.—Beds of marl had been long known to exist in the region around Charleston, South Carolina, and it had been used in a primitive way as a fertilizer for nearly half a century. These layers of marl, however, were in many instances overlaid by a stratum of rock, that had the appear-ance of ordinary limestone; and, in obtaining the marl, this rock, being in the way, was thrown aside as use-less. In 1867 the attention of Dr. St. Julien Kave-nal was drawn to this layer of rock, "marl-stones," which was unique in form, inasmuch as "it is in masses or nodules, varying from the size of a potato to several feet in diameter." The rounded shape was evidently " caused by the action of waves and currents " in the geological past. Dr. Kavenal analyzed these rounded stones, and discovered that they contained phosphate of lime from 55 to 60 per cent. Then followed discoveries of the rock in other localities, and the fact was also noticed that the rivers in the vicinity flowed over beds paved by this singular rock, " in a layer or sheet of cemented or tightly compacted nodules." Thus far it has been ascertained that, in South Carolina, an area underlaid with phosphate-rock, extends " seventy miles in length parallel to the coast, and with a maximum width of thirty miles." This phosphate-rock is deemed of immense value to the State and wherever the fertilizers derived from it are used. Phosphorus is an important factor in increasing the productiveness of the soil, and any substance that yields it becomes a fertilizer.

Almost immediately after Dr. Kavenal's discovery, measures were taken to utilize this phosphate-rock; experiments were made and factories established to prepare it by crushing and other processes. Immense amounts of the rock are thus put in form to be used by agriculturists, and are sent in every direction to enrich cultivated fields and gardens. The land-rock over the marl is mined, and the rivers are dredged and deprived of their bottoms of " cemented rocks " ; from the latter the State derives a revenue. We can not go into detail, but we may say that in 1884 there were in South Carolina alone more than forty companies, great and small, engaged in the work of preparing fertilizers from phosphate-rock, and that the latter is practically inexhaustible. (*Handbook of South Carolina, Mineral Resources of United States, 1882, and 1883–1884.*)

North Carolina Phosphates.—In connection with the large area of marl deposits in this State have recently (1884) been discovered, by special survey, nearly one hundred and fifty beds of phosphate rocks or nodules. The series of these as yet undeveloped deposits is in a belt of territory, fifteen to twenty miles wide, that extends from the Neuse River to the State line of South Carolina. This belt runs parallel with the coast, and from twenty to twenty-five miles inland, covering as far as known portions of eight counties.

Florida Phosphates.—This State has large deposits of marl, but as yet little utilized, and in addition phosphate-rock has been discovered. It has been found in the northwestern part of the State and in portions of four counties; deposits of the same rock are numerous in the central part, in the vicinity of Gainesville. In places these beds are from six to eight feet in thickness. (*Mineral Resources of the United States, 1883–1884, pp. 788, 793.*)

Maryland and Delaware Marl.—The former of these States has marl of the greensand variety in localities quite widely apart; having no commercial value, its use is confined to the neighborhood in which it is found. Delaware has also marl of the same kind in isolated and small beds, and is subject to the same conditions as to its use.

New Jersey Marl.—This State has vast deposits of greensand-marl, which occurs in beds or layers in a belt of territory running southwest from Raritan Bay to the Delaware River, about ninety miles in length, and an average width of perhaps eight. This fertilizer has been found adapted more or less for all crops grown in the State; it is used for ordinary products, for garden vegetables, fruit-trees, and vines of all classes—for grass-lands as well as for cereals. In some places it is dug by dredging-machines driven by steam, which, scooping it up very rapidly, drop it into the cars to be transported where needed. Marl is found in very limited quantities and in widely separated localities north of New Jersey, such as in small marshes that may supply local wants.

GYPSUM.

Lime under three forms is used as a fertilizer: The *first*, and by far the most extensive in its universality, is the lime derived from the common limestone by burning. This process is very simple, and the farmers under ordinary circumstances can prepare it themselves. The *second* form is phosphate rock, which is ground to powder, and, without being burned, applied to the soil. The *third* form is gypsum, the sulphate—sulphuric acid and lime—which, in being prepared for fertilizing purposes, is not burned, but ground fine. It may be mentioned that gypsum is sometimes a valuable building-stone, and that when it is burned it produces plaster-of-Paris—thus named from the city where it was first made. This form is known commercially as stucco, it being used in stucco-work in houses, as in cornices, and numberless forms of ornamentation. It has the peculiar property that, when moist and soft and correctly prepared, of setting, or becoming fixed in form within a few moments, when pressed into molds; hence the facility, for illustration, with which portions of cornices, etc., can be made alike and afterward easily joined together by the workman. This property of "setting" in plaster-of-Paris leads to its application in numberless instances; but to enter upon the detail of such is not within the scope of this book.

The uniting of sulphuric acid with lime results in a greater variety of compounds than either that of carbonic or phosphoric acids—those of the two latter being limited almost entirely to limestone and phosphate rock. Sulphuric acid and the carbonate of lime, united, produce several grades of gypsum, from the ordinary building-stone, of different degrees of hardness and durability, to alabaster of so fine a texture that from it are carved vases and mantel ornaments, etc. Alabaster is susceptible of a fine polish, and occurs in various shades of color, sometimes exceedingly white and translucent, and often yellow, red, or gray, and also " in transparent crystals or in crystalline masses," known in geology as selenite.

Gypsum; where found.—Gypsum occurs on the Atlantic slope, as far as known, in only two large deposits: one, in the central portion of the State of New York, and which extends at intervals through the State from Niagara to Oneida, and is in beds of great thickness—from it plaster for fertilizing purposes is manufactured at a number of localities along the line of deposits; the other, in the southwestern part of Virginia, in the valley of the North Branch of the Holston River, in the region where the village of Abingdon is situated. It is also found in a large deposit in Northeastern Alabama, in the valley of the Tennessee, within Jackson County. In Southwestern Louisiana, when boring for oil at a place some dozen miles from Lake Charles, a bed of gypsum, one hundred and forty-eight feet thick, was discovered; and in Texas it is reported to be in store along the head-waters of the Red River, in the northwestern portion of the State. Arkansas has gypsum in large quantities, and in this State are found beds of *alabaster*, said to be of most excellent quality and exceedingly white. Kansas has large deposits in the central portion of the State. Iowa (page 285) has immense quarries, but used almost entirely for building-stone, though a portion is made into plaster-of-Paris and for agricultural purposes. Ohio has several extensive beds of gypsum from which are manufactured, at the city of Sandusky, large quantities of plaster-of-Paris.

Michigan Gypsum.—This State has very valuable

deposits of this rock on Grand River, near the city of Grand Rapids, in the western part of the State, and, also, almost directly east from the latter city on or near the shore of Lake Huron, a few miles south of Alabaster Point. The indications are that these layers or beds of gypsum extend across the entire State between these two points. Some of these beds, as ascertained by borings, are found at various depths ranging to seven hundred feet below the surface. Michigan gypsum, when analyzed, gives in the main, sulphuric acid, 46; lime, 33; and water of crystallization, 21 per cent. Experiment shows that these elements enter largely into the composition of all cereals, clover, and other grasses; and into plants and vegetables, hence the immense value to agriculturists of the fertilizers derived from these beds of gypsum. In consequence, large establishments for preparing it are located at the quarries in the vicinity of Grand Rapids, while large quantities of the rock are sent into the State of Wisconsin, to be manufactured; and there are also exported large amounts from the eastern quarries on Lake Huron to factories in other States and to Canada, as the producers of fertilizers almost everywhere use more or less of prepared gypsum.

In the quarries near Grand Rapids are six beds of gypsum lying one above another, and over all are twenty feet of common earth. Between these beds are interspersed layers of soft shale, slate, and clay-slate, while the beds of gypsum are in thickness respectively eight, twelve, six and a half, eight and a half, nine and a half, and twelve and a half feet—in the aggregate fifty-seven feet of available rock, while the entire depth of these several layers of different kinds is ninety-eight feet. From these statements the reader may form a conception of the vastness of these deposits, which extend probably at the same thickness across the State from Lake Michigan to Lake Huron. "Thus far none but the two upper beds have been worked, and probably several generations will have succeeded one another before the necessity shall arise for resorting to the lower deposits for a supply of gypsum." (*Mineral Statistics of Michigan, 1881, p. 8.*)

The Workings.—The gypsum is varied in color, as white and rose-colored and mottled with gray. The principal mine at Grand Rapids is opened by three inclined shafts, extending from the base of the bluff to the bottom of the second layer of rock. The underground workings in 1881 comprised an area of sixteen acres, and of course now (1887) much more. The mine is free from dampness and is lighted with gas made within its limits. The rock is soft and easily drilled and blasted out with gunpowder. There are in the vicinity of Grand Rapids some half-dozen companies engaged in mining gypsum and preparing land-plaster and stucco; the amount of both classes annually sent to market is enormously great.

Rocky Mountain Gypsum. — Large deposits of gypsum have been discovered in Montana, and around the Black Hills of Dakota, though in limited quantities; but as yet these have been only partially developed, for neither land-plaster nor stucco has been brought into general use in that region. Utah has also an abundance of the mineral, which occurring in many localities has been utilized to supply domestic wants. Extensive beds of gypsum are found in the South and Middle Parks of Colorado, and along the base of the mountains, running east and west in a number of localities. At Colorado City is an establishment for the manufacture of plaster-of-Paris.

New Mexico has immense deposits in Rio Arriba, Socorro, Grant, and other counties. In the Sandia Mountains the natives make plaster-of-Paris in a primitive way, with which they whiten the interior of their adobe houses.

Arizona has likewise an abundance of gypsum, and widely diffused, ready to supply wants when needed. In Yavapai County it occurs in extensive beds of thin horizontal sheets; in Pinal County it is found in ledges where it is as white as snow.

Pacific Coast Gypsum.—Nevada is well supplied with deposits of gypsum, which is found in a number of localities, waiting to be utilized. California has extensive deposits of this mineral in Los Angeles County; these extend for twenty or more miles, and the gypsum appears of good quality. That found in Santa Barbara County is white in color, and resembles the Nova Scotia variety in texture. Gypsum has also been discovered in many places north of San Francisco, in the Coast Range, and in a number of other places within the State.

California has also in a number of localities the more delicate varieties, as alabaster, selenite, and satin-spar. The first-named is abundant in San Luis Obispo County, south of San Francisco, in the west slope of the Coast Range, near Aroyo Grande; selenite occurs in large slabs in Los Angeles County, and satin-spar in abundance in a number of places, as in Tulare and San Bernardino Counties.

The Importance of the Subject.—It is an inheritance of untold value to the people of the United States to have ample resources that enable them, not only to preserve, but to increase from year to year, the fertility of their soil. The American farmers, gardeners, and planters expended, in 1884, about twenty-seven million dollars in manufactured fertilizers—an immense sum, but not equal to the demand, for to this should be added the much larger aggregated value of the domestic manures, which the farmers themselves obtained on their own premises. Writers who have made the subject a study say that " comparison shows that only one seventeenth as much is returned to American soil, in proportion to the crops harvested, as in Germany," though the motto of every farmer ought to be " to return to the soil, each season, as much plant-food as the previous crop carried away."

Aided by the science of chemistry, the manufacturers of fertilizers are now able to use to advantage numberless substances that were once thrown aside as waste. At one time, bones of animals were sought, as the most prolific source of phosphorus, but now our immense deposits of phosphate rock on the South Atlantic slope supply all demands, and are also fast superseding the fertilizing material imported from Peru and the Pacific isles; in truth, we virtually import only one such material—potassium —from Germany; "the crude mineral, as obtained from the mines, is shipped for use as an ingredient in fertilizers." On the contrary, we are now exporting, as about 75 per cent of the phosphate rock, mined in the vicinity of Beaufort, South Carolina, is taken principally to England and Germany, while to the same countries is exported a very small proportion of that mined near Charleston. Thus we see that the numerous establishments throughout the Union for preparing fertilizers virtually derive their materials from our own resources.

XLV.

THE United States are peculiarly rich in the available food-resources of the ocean, they being within a convenient distance. In addition, the Americans have the right, by the law of nations, to fish outside three miles of the shore in the waters around Nova Scotia, Newfoundland, and Prince Edward Island; they have also, by treaty arrangement, the privilege of preserving or curing their fish on the shores of the same. Along the coast of the United States the ocean teems with fish, more or less at all seasons; and, at special times, certain classes swarm in these waters and are caught in immense numbers. About 130 miles southeast of Gloucester, Massachusetts, the great fishing-port of the Union, are the St. George's Shoals, where congregate at certain seasons multitudes of cod, to feed on the sea-weed or ocean-grass, as they do on the Banks of Newfoundland, the most celebrated and extensive fishing-grounds in the world. The cod is a rover in the ocean, somewhat like the buffalo on land: the one comes from the great depths at certain seasons to feed on the banks or shoals; the other at certain times to their pasturage in choice places on the plains.

Our Ocean Fishing Area.—Our own ocean fishing area commences off Eastern Maine, and, taking in the bays and sounds along the Atlantic coast, sweeps round Florida into the Gulf of Mexico, and along the north shore of the latter to the mouth of the Rio Grande. A similar area commences on the Pacific at San Diego, and passing up the coast includes the remarkable fishing-grounds of Columbia River, and still farther north those of Puget Sound, the Strait of Fuca, and thence to Alaska, whose southeastern and southern shores for more than a thousand miles teem with fish in unsurpassed numbers and of the best quality; thence round the Aleutian Isles into Behring Sea, where are found the fur-bearing seals at the Pribyloff Islands, and the salmon of the Yukon. This is only a brief outline of one of Nature's store-houses of food for the people at large, and all easily available for successful prosecution by our noble, hardy, and daring fishermen. About thirteen hundred species of fish are known to exist in North America, nearly all of which are found in the waters in and around the United States.

Discovery of the Cod.—The early navigators and explorers, both French and English, noticed the immense numbers of codfish that frequented the Banks of Newfoundland and the adjacent waters, and from that day to this the world has drawn largely upon these banks for supplies of food in the form of fish. It would seem that at the present time Americans capture a greater number of cod in these waters than do others; the quantity taken is enormous, and in a salted or cured form these fish, in part, furnish food for the people of the Union, besides the amount sent to Europe and elsewhere. The city of Gloucester sends out ships annually in different directions—to the Banks, to St. George's Shoals (of which the Americans have almost the monopoly), and along the northern coast—more than 500 fishing-vessels. At one season they go for cod; at another, fish in deep waters for halibut; at another, nearer the surface for herring and *mackerel.* The latter is a capricious fish; its habits are almost unknown to the most experienced fishermen. They will suddenly disappear from one point, and as suddenly appear at another; they are sometimes seen, in countless thousands, swimming near the surface of the water, but they may, as quick as a flash, dive down and be seen no more. Their ever-watchful enemies, the fishermen, cautiously approach, and, before they dive, throw out their nets and speedily draw them around the shoal, closing the nets underneath so as to scoop up the fish.

Utility of Ice.—In this connection it is fitting to notice the very great advantages the American people derive from the application of ice in preserving ocean food-fish in such manner that they can be sent in steamers along the coast, or on railways hundreds of miles inland to supply the daily markets in the cities, the fish meanwhile remaining fresh and wholesome. The same precautions are used in transporting the large numbers of fish caught in the Great Lakes. This ice-resource is becoming more and more appreciated and utilized in furnishing fresh food of various kinds, such as fish and meats, mutton, beef, etc., sent in refrigerator-cars, many hundreds of miles from the pasture-lands of the interior to the Eastern and other cities. In these cars, by means of a peculiar process, the *cold air is made dry ;* and the moisture, that arises from the partially melting ice, and which is so injurious to the meat, is neutralized, so that the latter is preserved fresh and in an unchanged condition. This application of ice, in connection with the power of steam, greatly enhances the value of our varied food-supplies, including fruits as well as meats, and from the necessity of the case its application in this respect will be increased indefinitely in the future.

The **Different Fishing** Localities.—Deep or ocean fishing is carried on almost entirely by the greater portion of the New England fishermen, while some also prosecute the in-shore fisheries along the coast. The deep-ocean fish of the North are deemed richer in flavor and general excellence than those of the South Atlantic; the latter are more migratory in their habits. In the bays and sounds, and near the shore along the Northern coast, some classes of fish, as the tautog, the black-fish, the sea-bass, etc., re-

32

main the year through, while in the spring and early
summer come the migratory shad and salmon, to run
up the rivers that are accessible to the ocean—the
salmon of to-day only in those of Maine and Nova
Scotia, and the shad in the remainder. During the
summer months, in addition to the usual fish along
the coast north of Delaware Bay, come the blue-fish
and the Spanish mackerel—both so desirable for the
table. The latter is deemed a delicacy: and, though
of southern origin, it appears to be gradually changing its habitat toward the north: this movement has
been noticed and recorded by naturalists. To supply these various fish for the markets of the cities
near the ocean, and also for those farther inland, requires the labor of large numbers of fishermen, who,
in prosecuting their occupation, move from station
to station as the fish appear off-shore.

The Shad.—The shad is the most important of
the fishes that ascend our Eastern rivers to spawn.
During the season, which commences in early spring
in the South and gradually extends North, they furnish a vast amount of delicious food, and, owing to improved facilities, both in preserving and transporting
them, they can be sent in a perfect condition far inland. In Long Island Sound and on the ocean outside, and farther south, in the bays and inlets of
North Carolina, come the menhaden, or white-fish,
or moss-bunker, swimming on the surface in untold
millions. They are caught in seines, and are used in
great numbers as a fertilizer, and also are pressed to

obtain oil. Though their flesh is somewhat oily, it is nutritious, sweet, and well-flavored, but their bones are so numerous as almost to preclude their use as a table-fish. However, an ingenious machine of recent invention has come to the rescue, as in a moment or two it takes out the bones, and leaves the flesh alone to be prepared, after the manner of French and Italian sardines, which they are said to rival in sweetness and flavor. (*Simmonds, p. 82.*)

Fish Products of the South Atlantic.—The Southern fisheries, in one respect, present features different from the Northern, as the former have no fisheries in the deep sea, but extensive ones nearer the shore on the ocean itself, and also in the adjacent sounds and inlets. When the fishing-season comes in the spring and early summer, the coast of North Carolina abounds in shad and also in herring. The latter are caught in seines in Albemarle and Pamlico Sounds, at the rate of two or three hundred thousand a day, and the shad at the rate of two or three thousand a catch. " Seven thousand seven hundred and fifty tons of herrings have been caught in a single season, and of shad fifteen hundred. . . . Steamers are at the wharves constantly loading with these fine fish, packed in ice, for New York and other Northern markets." (*Handbook of North Carolina, p. 334, and Simmonds, p. 60.*) In addition to these, large quantities are cured by means of salt, and inclosed in appropriate vessels, that they may be sent to all parts of the country during the year. Here seems, to human view, an unfailing source of nourishing food, obtained at the expense of only taking it out of the Atlantic.

The mullet, a favorite table-fish, also abounds in the waters just mentioned, but more abundantly in the inlets on the Florida coast. This State has a sea-coast of about a thousand miles, extending round into the Gulf of Mexico, " with numerous bays, sounds, and lagoons," and rivers that teem with food-fishes.

The Gulf of Mexico seems not inferior in propor-tion to its size to the Atlantic in the production of fish, all of which appear to have their habitat in southern or tropical waters. The north shore of the Gulf, amid its numerous bays and inlets, is specially prolific in fish, and where are seen the fishermen's primitive homes, oftentimes forming almost villages. The most prominent fish of this region is the *red-snapper*—thus named from its color—a moderately large fish and of fair flavor ; and also the crouper, a fish usually about the same size. Great numbers of these are caught, and find a ready sale in " New Or-leans and the cities around the Gulf, while they are also found occasionally in the markets as far north as New York." On this shore and amid the islands "fish and waterfowl abound in countless thousands," while there are an abundance of sea-turtle and crabs. "The water is thick with shoals of shrimp," which are prepared in enormous numbers for market ; here are likewise the finest varieties of fish. (*Louisiana Resources, p. 55.*)

The Pacific Fisheries.—From San Diego to the Strait of Fuca—about twelve hundred miles—the ordinary fish whose habitat is in that ocean are found in greater or less numbers along the coast; though, in the part of the ocean around Southern California, the best fish are wanting that are found in the corresponding latitudes in the Atlantic, notwithstanding in these waters there is an abundance of excellent fish of a different variety. Among these is the barraenda, nearly three feet long, lithe and shapely, and one of the best; and the "great ocean-pickerel," much brighter and clearer in color than his freshwater brother; and the red-fish or kelp, thus named because it frequents the sea-weed or kelp; and the rock-cod, one of the best table-fishes upon the coast: of this there are several varieties, and all rich in flavor. The Spanish mackerel also abounds—that is, a species, as it belongs to the same family, but is not equal in flavor to his Atlantic namesake. This fish is often two feet long, and weighs from eight to twelve pounds.

The most important fisheries belonging to the Pacific are, however, in the waters of Oregon and Washington Territory, and off their coasts, and that south of Alaska. This portion is stocked with the excellent classes of fish that are found in the Atlantic, on the same parallels of latitude, such as cod —three varieties—herring, halibut, and other kinds; while the salmon, in five varieties, visit the rivers: but as yet shad do not run up the rivers that flow

into the Pacific. It is reported that the Fish Commission intends making an effort to propagate the
shad in the waters of the Pacific.

The Salmon-Fishery. — The Columbia and its
tributaries rise, amid ice and snow, along mountain-
crests, and in consequence the waters of that great
river are remarkable for their clearness and coldness. Far up these streams, that prince of fishes,
the "spring silver" salmon, has chosen favorite
places to deposit its spawn. In immense numbers
they ascend the river, during the months from April
to August inclusive. The salmon can be captured
only at night, by nets stretched across the current
anywhere in the wide river. The water is very
deep, sometimes fifty feet, and wonderfully clear, so
that in the daytime the fish see the nets, and avoid
the danger, by either swimming above or below
them.

The canneries for salmon, so extensive in their
operations, are located at Astoria and in the vicinity.
The process is very interesting, because of its efficiency in retaining the flavor of the fish, to secure
which the work must be done within eight to twelve
hours. The fish are brought in at dawn or nearly so;
are immediately dressed and cleansed, care being
taken to remove the blood—for the salmon bleeds
profusely ; are cut into proper size, placed within
sealed cans, which are immersed by means of machinery in immense caldrons of boiling water, in
which the fish is thoroughly cooked ; the air forced

out through a pin-hole, which is closed by a drop of melted solder the moment the can leaves the water. Then, after being tested, that no imperfect can may leave the premises, the fish are ready for use. Ten thousand tons have been canned here in a single year, and as many more salted and put up in barrels. These salmon thus canned, or otherwise prepared, go over the civilized world—England taking about five thousand tons. Only the largest are caught, the meshes of the net being about eight inches in width, in which the fish become entangled by their gills. The enormous amount of salmon captured each year is gradually diminishing the numbers that come in from the ocean. Congress is presumed to have a care for the welfare of future generations as well as for the present one ; and it is due to the former that this system, so selfish and so destructive, should be abolished or properly regulated.

Puget Sound and Alaska.—The whole of the tide-waters of Puget Sound, from the entrance of the Strait of Fuca to its southern extremity, abound with food-fishes. These comprise the salmon, halibut, herring, and cod, with a number of minor varieties. The halibut, though apparently of Arctic origin, is found in the waters of the North Pacific ; it is a valuable food-fish, and is captured in large numbers by the Indians all along off the shores of British Columbia and Alaska. Twenty-five miles off the opening of the Strait of Fuca is a bank or shallow in the ocean, that is a favorite feeding-place for halibut, and where they congregate in great numbers. These fish are of unusual size, ranging in weight from seventy-five to three hundred pounds. The smoothness of the ocean aids very much in their capture. The great fisheries of the North Pacific are waiting for future development, though their products have already found their way to China and Japan, as well as to all parts of the Union.

In Alaska, toward the latter part of June, commences the run of salmon up the Yukon River. They come in millions, but because of the shortness of the season, and the roundabout distance to reach the river, this fishery has as yet been but little prosecuted. These fish are of the finest quality, both in size and flavor, and for this reason they are often characterized by fishermen as the "king salmon of Alaska." The waters along the southern shore of this Territory are specially prolific in several varieties of fish—the cod and halibut of unusual size, and the quality being very fine, and herrings also in untold numbers. " In some places the waters teem with fish to such an extent that they seem to be boiling. . . . The curing and canning of fish have already assumed large proportions." (*The Governor's Report, p. 5.*)

Fresh-Water Fish.—We have noticed thus far only the fish found in the ocean, and the two—the shad and the salmon—which annually visit certain rivers that are accessible from the ocean.

The Great Lakes on our northern frontier are

well stocked with fish of different varieties, many of which are valuable for food—of the latter class the white-fish being pre-eminent. It has properties very similar to those of the shad, as it possesses somewhat the delicacy and flavor of that popular fish; in the main it is equally a favorite, and is characterized as a fresh-water shad. It is captured in great numbers in the lakes above Ontario, and is sent long distances in refrigerator-cars to the inland cities, and is found often in the markets of those on the seaboard.

Fish-Culture.—The resource for fish-food, as contained in our numerous lakes and rivers throughout the Union, is worth to the nation many millions, if such benefits could be enumerated only in dollars and cents. This value is increasing from year to year, owing to the beneficial results derived from the recent introduction by the National as well as some of the State governments, and private associations, of the system of *fish-culture*, by means of which our lakes and rivers are not only stocked with the best classes of our own fish, but Europe and Asia are laid under contribution to furnish us specimens of their best varieties. Though this enterprise is only in its infancy, the results thus far attained indicate that the supplies of fish-food will continue increasing in proportion to the demand.

SHELL-FISH.

Shell-fish, such as oysters and clams, flourish in waters that are more or less salt; hence their favorite homes are at the mouths of rivers, and inlets or in bays or sounds that are intimately connected with the ocean, and where the agitation of the water is not violent, it being moved mostly by the flow and ebb of the tide. They always breed and multiply in those places where the bottom is a soft alluvial deposit, in which they can burrow and find food. The area where this class of fish abounds extends so as to take in for the most part the adjoining sounds, bays, and inlets along the entire coast of the United States, on both oceans, and on the Gulf of Mexico. There are, of course, unsheltered spaces intervening, where shell-fish can not exist, because of the rapid motion of the water, which renders the bottom too hard for them to burrow.

Oysters and Clams.—The oyster is a very sensi-

Fig. 13.—Oysters.

tive creature in respect to what it feeds upon, and in consequence the flavor of its flesh depends so much upon the quality of its food that the peculiar characteristics that any class may possess are designated by the name of the locality in which they were originally found, rather than assigned to a separate species. It thrives in deeper water than the clam, and is taken up from its shallow bed on the bottom by peculiarly shaped grappling-irons at the ends of two long handles, which are worked by the oyster-man. The variety of shell-fish known as the scallop is found in abundance in the waters along the coast of the United States.

The clam is deemed in value second only to the oyster; of it there are two or three distinctive varieties, as the round, the long-necked, etc. Clam-beds occur in sheltered places that are left bare by the receding tide, and these are found in different localities from Cape Cod to Florida. Clams of all kinds burrow so deep in the soft or muddy bottom that they are taken from their hiding-places only at low tide.

Lobsters.—The lobster is more refined in its nature than either the oyster or the clam; it prefers clear water and a clean bottom, where its food is obtained. Its favorite home is in deep waters in the bays and sounds, or a short distance from the shore in the ocean itself. The lobster is captured by means of boxes called pots, which are anchored on the bottom, and in which is placed bait, and when

the lobster enters the box to obtain the choice mor-
sel its destiny is sealed. Lobsters abound in great
numbers off the coast of New England a short distance from the land; they are also found in the deep
waters within the bays along that coast and in Long
Island Sound. They are in great demand as a delicacy rather than for substantial food. Within recent
years they have been taken in such multitudes as to
sensibly diminish their numbers, and, if measures are
not soon taken by the Government to regulate their
capture, the descendants of those who are now, for
the sake of gain, destroying them, will scarcely have
any. The great mass of those taken are canned, and
sent abroad mostly to England; they are also sent to
the cities in the interior of our own country. Says
an authority on the subject: "A few years ago it
was not uncommon to catch lobsters weighing from
ten to twenty pounds; now the average is from *three
to six pounds.*" This fact plainly indicates what will
soon be the result.

Oysters in the Sound.—Passing from New England, we find Long Island Sound lined along its shores,
from end to end, with beds of native oysters wherever the conditions are favorable, or where they can
be cultivated. These beds occur sometimes in the
deeper water, and have been often discovered where
least expected. Almost side by side with the oyster-
beds are frequently found those of the clams; the
latter being more prolific and natives, as they are never
planted. Oysters are often brought in vessels from

Chesapeake Bay to the Sound and dropped overboard in suitable places; there they are left to themselves to obtain food, and grow into good condition for the table. Such oysters oftentimes become acclimatized and developed as finely as the native, whose place they were brought to fill when the latter became exhausted in supplying the demand. These beds are not confined to the Sound itself, but they extend around New York harbor, and in the small bays and inlets along the Jersey shore and vicinity; counting the shores of inlets which are from twenty to thirty miles in length, the whole distance is crowded more or less with beds of oysters and of clams. Within recent years successful efforts have been made, specially on the north shore of the Sound, to cultivate more carefully the native oyster, and it is found that, when properly cared for, they become large and luscious. Here is a vast and well-defined resource, the outcome of which is increasing every year in proportion to the demand. An estimate can be made of the amount of labor and capital invested in this means of supplying one class of food to the people, when, in the vicinity of New York city alone, are engaged in this business more than eight hundred seamen, and about one hundred and seventy-five sailing-vessels of all grades.

Oysters in Chesapeake Bay. — Farther south come this noble bay and its numerous inlets and mouths of rivers, wherein the oyster flourishes better than anywhere else in the Union. This remarkable water—about two hundred miles in length and from four to forty in width—is stocked with fish from Hampton Roads to the mouth of the Susquehanna at its head, while its bottom may be said to be covered with beds of oysters unquestionably the finest in the world. In all the employments connected with this great store-house of shell-fish food, it is estimated, are engaged about thirty thousand persons. On this bay or in connection with it are two important points in this trade in marine food —Norfolk, Virginia, and Baltimore, Maryland. Says an English authority : " Baltimore ships raw oysters to South America, California, and Australia, and besides to all parts of Europe ; and the demand will steadily increase as they become better known, from the fact that Chesapeake oysters, like canvas-back ducks, owe their superior flavor to the food obtained on their feeding-grounds." (*Simmonds, Wealth of the Ocean, p. 144.*) In Baltimore are also numerous large establishments engaged, during the winter and early spring months, in canning or pickling oysters for consumption in the United States and in foreign countries. The thousands thus employed are likewise engaged during the summer and the autumn in canning small fruits and vegetables, which are so abundantly produced in the gardens in the region for a distance round the city. Oysters occur still farther south in the inlets and bays along the coast from Virginia to Florida, but have been thus far utilized only for domestic use, in the immediate vicinity where found.

Oysters in the **Gulf.**—On the north shore of the Gulf of Mexico, where are numerous inlets and small bays, oysters abound. They are specially plentiful and of fine quality in the Bay of Mobile, and in the adjoining waters, they being cultivated on quite a large scale. (*State of Alabama, p. 116.*) In Louisiana are extensive oyster-beds along the southern coast and bayous; the size and flavor of these oysters are claimed to be unsurpassed. Some of them "are so large that they are not merchantable to saloon-keepers, who buy them by the barrel and sell by the dozen. The canning of the Gulf shrimp is attracting attention, and the business is increasing." (*Louisiana Resources, pp. 7, 8.*)

Oysters and Clams in the **Northwest.**—Oysters and clams abound in great abundance in the Strait of Fuca, and in Puget Sound—the Chesapeake of the Northwest. The native oyster is quite small, but of good flavor; those of Shoalwater Bay, on the coast, are plentiful, and find a ready market in San Francisco. The Eastern oyster has been introduced sufficiently to prove that it can be successfully cultivated. Here is a wide field for enterprise. "Clams of several varieties abound, and range in size from one to ten inches, and from one ounce to ten pounds in weight." It is hoped the United States Fish Commission will soon introduce the Eastern lobster to these waters, and, if possible, the shad and the mackerel. (*Report of Governor, 1886, p. 37.*)

Green-Turtle Fishing.—The feeding-grounds of

the turtle are at intervals around Florida, from Fernandina on the Atlantic, to St. Marks on the north shore of the Gulf of Mexico. The prominent places of this fishing industry are Key West and Cedar Keys, the latter being near shoals in the Gulf that are extensive feeding-grounds for the turtle; these extend from St. Marks south along the coast to Key West. The turtles are captured in nets with large meshes, in which they become entangled; as soon as caught they are shipped to market, the object being to send them when alive. (*Florida Times-Union, 1885.*)

Sponge-Fishery.—In 1852 it was discovered that in the waters around a portion of Florida as fine sponges were to be found as are produced in the Mediterranean. At Key West and in the vicinity was commenced the industry of obtaining sponges and of preparing them for use; the business increased from year to year, as the fine quality of the sponges became better known. Meanwhile the localities where they were first discovered became nearly exhausted, but a much larger area was found in the Gulf, and in 1870 Appalachicola became actively engaged in fishing for sponges. In the Mediterranean the fishermen dive to the bottom for them; but the Americans use a two-pronged hook at the end of a slender pole some forty feet long, according to the depth of the water, meantime using what they call a " water-glass "— which is like a bucket with a glass bottom—that enables them to see the sponge. With these hooks the sponges are disengaged from the bottom and drawn up, and are spread upon the deck of the sloop in the sunshine and left to die. Afterward they are freed of a large portion of the glutinous matter that is connected with the fiber of the creature, then they are soaked for about a week in salt-water, which process loosens the skin or covering that had become dry and tough under the sunshine. It is now beaten, squeezed, and made partially clean, and taken to the packing-house where it is thoroughly cleansed, trimmed, and made ready for use. The work is more successfully prosecuted during the summer, and only when the water is both calm and clear. It has been discovered recently that with proper care the sponge can be cultivated to almost any extent.

XLVI.

WE possess virtually an exclusive but unique re-source of the ocean in the fur-bearing seals, whose summer resort is on the Pribyloff Islands—the principal ones being St. Paul and St. George—in Behring Sea, within the boundaries of Alaska. They come to these islands only at their breeding-season, which extends from about the first of May till near the middle of October; the remainder of the year they are supposed to roam over the North Pacific and the adjacent portions of the Arctic Ocean in search of their usual food—small fishes.

These seals have characteristics that no others have in the length and peculiarly fine texture of their fur, which is said to be the only fur that is uninjured by dyeing. The males are allowed by law to be captured from about the first of May till August, when they begin to shed their coats, and the fur becomes comparatively worthless. The skins are partially cured by means of salt and sent to San Francisco, thence for the most part to England. In London are the chief establishments in the world for dyeing and preparing the fur on these carefully tanned skins. The great beauty of the finished fur is attained by a process which is a secret to the outside world. It requires from twelve to eighteen dyeings and manipulations, to secure the exquisitely beautiful gloss of the bronze or the jet-black fur. The labor pays, as it produces the material for that comfortable article of dress, the seal-skin cloak worn by ladies. Fur-seal skins are also prepared to some extent in Albany, New York.

Efforts to preserve the Seals.—Avaricious hunters—white men and Indians—were rapidly destroying these seals, and even killing them at a season when the fur was almost valueless, when Alaska passed into the possession of the United States in 1867, at the price of $7,200,000. The National Government at once took measures for regulating the capture of the seals. A company, meanwhile, was formed and chartered to prosecute this seal-fishery, and which pays to the Treasury for the privilege a yearly rental of about $300,000, thus producing a revenue of *four per cent* on the original cost of Alaska. At first the number allowed to be taken was limited to 100,000 a year, that being deemed sufficient to supply the wants of the world; since then the number has been increased to 125,000. A United States revenue-cutter is always on hand to enforce the laws, and under these judicious regulations the seals are gradually increasing their numbers.

Peculiarities of this Seal.—This remarkable sea-animal has singular habits, of which a brief notice may interest the reader. The narrow beaches of the

FIG. 14.—Fur-bearing Seal of Alaska.

Pribyloff Islands are covered with large bowlders, detached from one another by irregular spaces that serve also as passage-ways. Instinct leads these seals to congregate here, during the breeding-season, in many myriads. The males come first, and select each one his district or bowlder, and take possession peaceably, if there are no rivals; but, if there are, they fight it out, and the one that masters holds the coveted bowlder, without further interference on the part of the one vanquished, or of others. Each one's district is bounded by the irregular spaces between the rocks: but these spaces or pathways are respected as neutral ground, and within them the bitterest enemies peaceably pass and repass one another; but, let one infringe upon the domain of the other, and at once a fight commences in order to eject the intruder.

Matters being thus arranged, they all await the coming of the females, on whose arrival commences a series of miscellaneous fights, but for another purpose. Each male strives to secure as many females as possible under his protection, and quiet does not prevail until these domestic arrangements are made, and which, after the contest is ended, are seldom disturbed. The male defends his family to the utmost, and guards the young when they are helpless, and is so diligent in his paternal duties that, during the season, he does not even go into the water to obtain food, but lives upon the fat which he has accumulated within the previous eight or nine months; but the female goes regularly into the water to feed, and as regularly returns to nurse her young. When the latter are old enough to seek their own food, parents and all leave the islands until the next season. It is thought by naturalists that this kind of seal does not attain its full size before it is five or six years old. These heads of families, both male and female, are exempt from capture. But there is another class that become victims: the latter are the males that do not known as " bachelors." These unfortunates congregate by themselves, and when, as far as can be judged, at the right age, they become the legitimate prey of the hunters, and thus furnish their skins and fur to warm and decorate the ladies.

In order to perpetuate the race, a United States law forbids the killing of the females. Formerly the latter were killed indiscriminately with the males, and the fur-bearing seal, owing to the greed of these hunters, was fast becoming extinct ; but the law came to the rescue, and which, as occasion requires, is enforced by a United States revenue-cutter.

WILD GAME.

At this day, in the cultivated portions of the Union, wild game is of no special value, unless it may be said of the varieties of birds of passage; but in early times, in these very regions, the game in the forests was of great importance to the original settler, as in a great measure he depended upon it for a large amount of the daily food for his family. So great was this demand in the early settlements that men engaged in hunting, in order to supply provisions for the people, in the same manner as men of to-day are engaged in fishing. Neither is it proper to depreciate this resource, because only in the days of the colonists, or of pioneer settlers, was it largely drawn upon, any more than we should hesitate to acknowledge the benefits conferred upon our fathers by the use of coal, though in the process it was itself consumed. The colony of Virginia, in the winter of 1607, was saved from starvation, or nearly so, by the numbers of wild ducks and geese which they captured on the lower Chesapeake.

Wild Animals and Birds.—It is to be noted as a great advantage that in the forests of the United States were found no such large and dangerous animals as the lion and the tiger, but on the contrary those of a harmless type, and suitable for food, as the deer, moose, or elk, of the East and North, and in the West the buffalo, the deer, and the antelope. The smaller food-animals, such as rabbits and squirrels, were found in multitudes. Of wild-fowl there were two classes: the aquatic, including the varieties of ducks and geese; while inland were the quail or partridge, the pigeon, and many of lesser note, and the prairie-hen or American grouse of the West.

The Turkey.—More than all in value, the turkey was found in large flocks, and ranged extensively in all the timber-lands of the country East as well as West. The latter is indigenous to America, and it adds immensely to its value that it was capable of being domesticated, which merit none of the others mentioned possessed. It takes the highest rank as a table delicacy, both in the United States and in Europe. John Cabot and his son Sebastian, on their first voyage to America (1497), were presented by the Indians with a male turkey and two hens. These were taken to Bristol, England, and from them are descended the turkeys of Europe. Had this voyage conferred no other benefit upon Europe than the introduction of this valuable bird, its expenses would have been more than amply repaid.

That wild game is still valued, especially by the settlers in the West, we incidentally learn from the notice it receives in some of the reports of Territorial Governors and in some of the new States : " Wild game abounds, and its wanton slaughter is prohibited, that it may be preserved for the use of the inhabitants. . . . This Territory has about one hundred and twenty-five species of birds, and from its central location may be taken as a criterion of the others adjoining. The wild turkey abounds in all the mountains of the Territory—specimens very often weighing from twenty to twenty-five pounds." (*Governor's Report, Wyoming, 1885.*)

" In all the valleys of the State [*Oregon*] abound deer, pheasants, grouse, quail, snipe—the last four of unusual size. . . . In the mountains, deer, elk, and antelope, in great numbers ; in autumn wild geese and ducks swarm along the water-courses. . . . The valleys and hills are grass-grown, and are alive with grouse and snipe, sage-hens, and prairie-chickens." [*Utah.*] " The forests abound with deer, grouse, and pheasants. . . . Elk, deer, and antelope are still abundant." (*Washington and Dakota Territories.*)

The Canvas-back Duck.—The food-animals of the forest, large and small, having thus completed their mission, virtually disappear from the eastern, cultivated portions of the Union ; of the game-birds, the turkey was domesticated, but those of passage still remained. The latter are the ducks and geese, in numerous varieties, that yearly frequent the bays, inlets, and sounds along the Atlantic coast. There is not in the world a water so prolific in birds of game as Chesapeake Bay, while farther south, in Albemarle and Pamlico Sounds, the same kinds are found in multitudes; but the Chesapeake is the favorite feeding-ground of the famed canvas-back duck— the latter found only within the boundaries of the United States. Says Dr. Lewis, an authority in respect to the birds of the Chesapeake: " All species of wild-fowl come here in countless myriads; and it is really necessary to visit the region in order to form a just idea of the wonderful multitudes, and numerous varieties of the ducks alone, that darken these waters, and hover in interminable flocks over these famous feeding-grounds." By far the most valuable of these birds is the celebrated canvas-back duck; it frequents the shallows where the water may be from six to ten feet in depth, and diving down finds on the bottom its favorite food, a species of *wild celery*, said to be found nowhere else ; it has a grass-like blade, and a white and tender root, whose peculiar flavor, being imparted to the flesh of the duck, renders it a table delicacy unsurpassed by that of any other bird.

Migration **of Ducks** and **Geese.**—It is an interesting fact, in relation to these ducks and geese, that they go sufficiently early in the spring to spend the latter portion of that season and the entire summer on the banks of the Yukon River, about 63° north latitude, in distant Alaska. The instinct of these birds is marvelous! The winged columns, as regular as battalions of soldiers, move from the Chesapeake and other bays and sounds along the South Atlantic, and on the north shore of the Gulf of Mexico, in a general northwest direction, till they come in contact with the Rocky Mountains, thence up their eastern base, and finally through their depressions, to the head-streams of the Yukon, and down them to the main river. There comes also another column, whose destination is the same, along the western base of the Sierra Nevada from the Gulf of California, and other bays and inlets of the Pacific. Along both banks of the Yukon, and for hundreds of miles, amid the sedges that extend back from the stream, they build their nests and rear their young, meanwhile feeding on the many varieties of berries that are produced in that region in immense profusion. (See page 458.) When the young are able to fly and ready for the journey, the entire number set out, and, never making a mistake, return by the same route to their winter homes and feeding-grounds.

Game-Food in Virginia and Minnesota.—Norfolk, Virginia, because of its location and facilities for transporting freight, has become a center of the trade in the game derived from the Chesapeake, and also from Albemarle and Pamlico Sounds. During the latter portion of autumn and through the winter rapid steamers and railways distribute all classes of this game wherever demanded. The mission of these wild ducks and geese is by no means limited in their furnishing food from the Chesapeake; for, on their journey in the spring to the Yukon, after flying some six or seven hundred miles, they must rest, and instinct leads them to alight where they can obtain food, amid the lakes of Minnesota and the adjacent States. Here they are also hunted for food, until they resume their journey, and when they arrive at the Yukon they are hunted again for the same reason by the Indians. However, in spite of these annoyances, they manage to become exceedingly fat, so that they can scarcely fly, because of their feeding on nutritious berries. On their way back, in the autumn, they again alight to rest among the same lakes where the previous spring they had stopped, and again to be harassed by the hunters. Thus, nearly the year through, these poor birds are furnishing food for man, and are free from their enemies only when, on their passage to and fro, they are out of gunshot, by being too high in the air.

It is well known that the wanton destruction of these birds on the Chesapeake has been gradually diminishing their numbers. They surely ought to be properly protected by legislative enactments, not only on the Chesapeake, but on the lakes of the Northwest, and perhaps it will yet be found necessary to extend similar protection to the banks of the Yukon.

XLVII.

WE of to-day do not appreciate so vividly as did our fathers the value of one natural resource—that of water-power—so important, when sufficient in quantity, because of its comparatively small expense. Until the era of steam, this was almost the only power used for driving machinery, such as mills for grinding grain, or for sawing and preparing lumber, also for manufacturing of various kinds; and to-day, in the interior highlands of the country, are almost innumerable small mills and factories of different kinds driven by water-power. In portions of the Union, especially where the winds from off the ocean were available, windmills were often used, and are yet in the interior in many portions of the land, usually for pumping water from wells. Many streams, however, have outlasted their usefulness, often from the lack of water caused by the diminution of the forests amid which are their head-streams, or they have been superseded by the use of steam. Yet there are in the Union numbers of rivers that "will run on forever" in doing good service for man.

The Merrimac **and** the Connecticut. — There trickle down on the southeast side of the White Mountains innumerable little streams that unite and form a clear, bright, and rapid rivulet, which is joined from time to time by other streamlets creeping out from ravines and valleys, until it becomes a small river, under the name of the Pemigewasset. Farther on its way, after being joined by others, it takes the name of the *Merrimac.* Who would imagine that this modest, this beautiful little river was really, in one respect, the most remarkable on the globe? Within the many cities that line its banks it drives more mills, and of greater extent for manufacturing purposes, than any other river known, in proportion to its length and amount of its waters. It confers benefits directly upon tens of thousands by being the occasion of giving them employment, and indirectly upon hundreds of thousands more. The *Connecticut* carries a much larger body of water, and for a much greater distance, than the Merrimac, and it claims a similar honor in being useful, on account of the numerous mills it drives, and of the wealthy manufacturing cities that have sprung into existence with wonderful rapidity along its middle portion, even within the third of a century. These two prominent rivers belong to New England, but all through the Middle States are numberless water-powers, though none are, or can be, utilized to so great an extent.

Other **Water-Powers.**—The Alleghanies on both

their eastern and western sides have head-streams, with their waterfalls and power: on the Atlantic slope are prominent the falls of the James in Virginia; on the southern end of these mountains, numerous streams, furnishing immense water-power, take their rise and course their way to the Gulf, as the Coosa and the Chattahoochee, and others. The most remarkable series of water-powers in that section, and which take their rise in the Alleghanies, are in the Tennessee, with its large volume of water, much more than that in the Merrimac and Connecticut combined, or in the Ohio. The river descends for thirty miles over a series of shoals called the "Muscles" (from that fresh-water shell-fish), thus "creating an amount of power greater than is to be found on the continent within the same compass." The National Government is constructing round the shoals a canal, to be finished in 1889, which, it is estimated, "will afford power enough to turn machinery" to an untold amount, and that without impairing its efficiency for the purposes of navigation. The water-power of the upper Mississippi is remarkable for its greatness, and for the benefits it has already conferred upon the people. In this respect may be cited the advantages which the cities of St. Paul and Minneapolis have derived from the Falls of St. Anthony.

LAND RESOURCES OF THE UNITED STATES.

Political economists, especially those of Europe, in endeavoring to account for the unprecedented material progress of the American people, take the ground that the chief basis upon which they secured success was the possession of almost a continent of cheap and fertile lands. It is well known that the territory of the Union, taken as a whole, is remarkable, not only for the richness of its soil, but also for a climate that fosters agricultural products. By means of Nature's highways, which, when needed, have been supplemented by railroads, the United States have unusually fine facilities for communication between different portions of their large domain. There is another reason, though not fully appreciated by these writers, as this land of pure springs and crystal brooks, of fertile soil, and often with minerals underneath, would have remained unutilized, had it not been for the unwonted *industry and energy* of the people. They appeared to receive a new inspiration of hope for themselves and their children when they became a Nation, and were freed from the trammels of foreign authority, and thus found relief from numerous drawbacks to their future prosperity; and the whole Nation under this new impulse bounded forward, as one man, to possess and subdue this goodly land.

Mistakes as to the Soil, etc., of the Great Plains. —The more early writers, tourists, and explorers often indulged in discouraging statements in respect to the fertility of portions of the great Western plains. These gloomy forebodings for the most part have vanished, since it has been practically demonstrated that nearly all these plains are capable of successful cultivation or of pasturage. It is found that some portions of these regions, which, hitherto, have been deemed unfertile if not absolutely barren, possess an alluvial soil, the *débris* of former ages, that is susceptible of producing crops abundantly by means of irrigation. (See page 389.) The great majority of such districts are so situated that they can draw upon the inexhaustible reservoirs of water derived from the melting snows in the mountains, and from the numerous streams that flow from the latter. Practical irrigation in the regions mentioned is as yet in its first stages, but the indications are that it will lead in the end to the successful cultivation of vast areas of the public lands now unoccupied, and this resource of the Union will be rendered still more available for the people at large.

Rainfall; its Gradual Increase.—A very striking feature exists in relation to the amount of rain that falls west of the one hundredth meridian, and also in respect to the time of year it occurs. As found by recorded observations, the annual rainfall is greater in the more eastern portions of the Union than on the plains of the wheat region of the Northwest; yet in the former the far greater portion of the rain comes in the winter and the early spring, before it can be available for the crops, while in the latter the heaviest rainfall occurs in May and June, when every drop is available. For illustration, " the rainfall, as shown by observations at the United States Signal Station, at Huron, South Dakota, has steadily increased, since the first sod in the vicinity was turned over." A dozen years before 1886 the lands that are now regarded as among the richest in that section were condemned as being unfit for cultivation. The black soil in this region runs from a few inches to from three to five feet in depth, every square foot of which is more rich than the richest in New England. (*Condensed from Dakota, pp. 12–21.*)

This increased rainfall seems to be the result wherever the land is brought under cultivation. " The yearly extension of the rain belt westward has been very apparent during the past few years, and of which due advantage has been taken by extreme Western settlers. Lands which five years ago (1879) had only a scanty covering of buffalo-grass, the soil baked by the hot suns of summer, are now sending up a growth of blue-stem and other strong native grasses, which will shade the soil. This is due to increased rains." In Nebraska are lands, three hundred miles west of the Missouri River, that have been hitherto regarded as worthless for agricultural purposes, but which to-day (1887) are producing fine crops of wheat, oats, rye, barley, and Indian corn.

The Reason of the Increased Rain.—The question

34

tion may be asked, What is the cause of this increase of moisture? The most reasonable solution is that rain follows the plow; as the soil is turned over, the rain-water, instead of running off as formerly from the unbroken prairie, penetrates the soil and is there held until gradually given back to the atmosphere by evaporation. Prof. Aughey, of the State University of Nebraska, has shown by a series of experiments that equal areas of the surface-soil, and of equal depth within the earth, and which were taken up immediately after a rain, from a plowed field and also from the unbroken prairie a few yards distant, the former retained *eight ninths* more water than the latter. The equal quantities of earth thus taken up were weighed, and, after being subjected to the same process of drying in order to evaporate their moisture, were again weighed, and the difference ascertained.

The yearly increase of rainfall is extending westward wherever the prairie is brought under cultivation, and intelligent persons on the ground, and who have studied the subject, believe that the entire slope from the base of the Rocky Mountains eastward will yet be used for pasturage or for agricultural purposes; in confirmation of which belief is adduced the fact that along the route once traveled by thousands to Pike's Peak, over what then appeared as a desert of sand, cactus, and sage-brush, and which Horace Greeley, who in 1859 passed over it for ninety miles, characterized as "the acme of barrenness and desolation," is now a farming and pastoral region. Moreover, this whole territory is underlaid by an abundant supply of pure, fresh water that can be reached by ordinary wells or by artesian. (*Nebraska Resources, 1885, pp. 24, 25.*) The water in these underground reservoirs that can be tapped by wells will in the future be indefinitely increased in quantity, as the surface becomes cultivated, thus giving freer access for the water to penetrate the earth. We have seen, by Prof. Aughey's experiments, that, from the matted surface of the unbroken prairie, *eight ninths* of the rain-water ran off, and only one ninth was retained.

These results and experiments in relation to the increased rainfall in certain regions, and in others the facilities for the necessary irrigation in order to produce crops abundantly, are but the forerunners of still greater agricultural and pastoral success and blessing to coming generations.

How Homesteads are obtained.—It is fitting, in closing a narrative of the Natural Resources of the Union, to show under what conditions portions of one of these, the unoccupied lands, may become available to those who desire to own a farm and a home. That beneficent act, the homestead law, after being bitterly opposed for a number of years in Congress, finally was passed, and, receiving the signature of Abraham Lincoln, went into effect on January 1, 1863. This law provides that any settler twenty-one years of age, male or female, the head of a family, "on payment of ten dollars"—to cover expenses of survey—"he or she shall thereupon be permitted to enter a quarter-section of unoccupied land" (one hundred and sixty acres). It is also provided that persons of foreign birth may enter homesteads, "provided the immigrant be a citizen of the United States, or has declared his intention to become such." In order to aid the worthy landless still further, the law provides that "this homestead shall not in any event become liable to the satisfaction of any debt or debts contracted prior to the issuing of the patent therefor." If the settler occupies the land in *good faith*, and improves and cultivates it for *five years*, on evidence of which compliance with the law, the United States Government gives him a *title in fee;* previous to the giving of this title, the homestead is *free* from taxation. The law makes provision also that, in case of the death of the settler before the expiration of the five years, the homestead is secured to his or her heirs.

Land-Grants and Railroads.—In order to facilitate access to these unoccupied lands and make them available for settlers, the National Government, in connection with the homestead law, made grants of land to corporations, which were pledged under certain conditions to build railways through these lands and across the continent. Three such roads have been built under this provision, and have thus aided in extending a line of settlements from the Mississippi to the Pacific. Where the roads passed through States that had public lands within their boundaries, the grants were ten miles wide on either side of the road, but within the Territories they were twenty. Owing to the nearness of the railways, the public lands within these grants were estimated at *twice* the money-value of those outside; hence, within these, the settler received as a homestead only eighty acres, and the land to be sold was held at two dollars and a half an acre. By this arrangement, neither the Government, nor the people whom it represents, *lost anything* by grants of lands to these railways—a fact but little noticed, though worthy of remembrance. Homesteads are also attainable under the conditions of the timber-culture act, and under the pre-emption law; the latter gives the first settler, in preference to others, the right to purchase the land that he is occupying. The latter law has also been of great benefit to the settlers of limited means. The immense advantages derived directly from the railroads that were built by means of funds based financially on these land-grants are not limited to the homesteaders alone, but also include all those who have settled within the grants, and purchased their farms from the railway corporations, as well as indirectly to the Nation at large.

The Government surveys the land and divides it into sections, or square miles, each containing 640 acres, and also into quarter-sections, of 160, and eighths of 80 acres. A township constitutes thirty-six sections or square miles, and for the support of schools, when the Territory becomes a State, the *first* and the *thirty-sixth* sections are reserved from sale or gift. If the land is within a railroad-grant the remaining thirty-four sections are divided evenly between the railroad and the Government. The railway corporations sell their lands on very liberal terms; as a rule, the payments being so arranged that the prudent and industrious purchaser can in a few years own his farm free from debt.

Results of the Homestead Law.—From the annual reports of the Secretary of the Interior we obtain a partial view of the results secured under this beneficent law, during twenty-three years (1863–1886). There were given in homestead-grants in those years, in round numbers, 112,000,000 acres, equal to 175,000 square miles. In the ordinary enumeration of the area of the States are included the surface area of the rivers and lakes and the ocean shore-line three miles out; but these areas are not reckoned in the surveys of land for homesteads, and therefore, in making a proper comparison of areas, such surface should be deducted. The area thus occupied in free farms, during these twenty-three years, is equal to the combined area of all the New England States, of New York, of Pennsylvania, and of Maryland, after deducting from the latter their water surface, estimated at 10,000 square miles. In respect to the amount of land given for each homestead—those within the railroad-grants being 80 acres, and those without 160—the average rate is, according to the

Interior Department, 120 acres; this gives 933,333 households on free farms, and that gives, on an average, five persons to each household, a population in round numbers of about 5,000,000 living on homestead lands. There are also, in addition, large numbers of citizens dwelling within the same territory who have been induced to migrate thither because of the facilities afforded by the railways, and who have purchased lands either from the Government or from the corporations of the railroads. Let it be remembered that to-day (1887), on these original homestead lands, are numerous villages and cities, ranging in population from 100 up to 12,000. Such is the outgrowth of the humane and enlightened statesmanship that enacted the homestead law, and in that connection projected railways to penetrate these unoccupied and remarkably fertile lands, which were then useless to the people, because of their distance and inaccessibility. The object of the National Government was to bring these lands within reach of the older and more densely populated portions of the Union; to afford an opportunity, for those who wished, to secure farms and homes; and at the same time, also, to promote the growth across the continent of settlements composed of industrious and moral citizens. There is still an abundance of more or less fertile public lands waiting to be occupied. They may not all lie directly upon the three main railways to the Pacific, or on their connections, of which there are already about seven. There is scarcely a doubt that a few years will see that entire region accessible by means of branch railroads, especially if within it are formed settlements. It is, however, essential for those who desire homesteads to go where the public lands *are located*, as they are not to be had in the vicinity of the cities.

History records no legislative enactment equal to the homestead law in its humane provisions, and in the judicious introduction of proper means to make these unoccupied lands available for the landless. This law, in its effect, has already conferred on millions of men, women, and children numberless substantial benefits, lasting in their nature, and destined to increase in importance from generation to generation.

XLVIII.

THAT the reader may have a conception of the continuous industrial progress in the United States, the following tabulated summary is given — this progress being based alone upon the use or development of their *natural resources*. To enable the reader to make estimates for the future, by an application of the rule of proportion, the output is given for three years. The chief authorities consulted are the statistics of our mineral resources, as published from year to year by the Interior Department at the city of Washington. It is a great pleasure for the author to acknowledge the kindness and the labor of Professor David T. Day, of the Interior Department, for the condensed statements in the column under 1886, derived as they are in advance from the forthcoming volume of United States mineral resources for that year. This summary includes the *output of our own resources ;* for instance, it sometimes happens that a manufactured article is made from foreign or imported material combined with native American —the latter only is given. This summary comprises but a portion of the output of our resources, as immense quantities are evidently not fully reported.

EXPLANATIONS.—L. t. = long tons (2,240 lbs.); s. t. = short tons (2,000 lbs.); oz. = ounces; bu. = bushels; v. = value; est. = estimated; bar. = barrels.

	1884.	1885.	1886.
METALS.			
Iron-ores, l. t	8,200,000	7,600,000	10,000,000
Pig-iron, l. t	4,097,868	4,044,525	5,683,329
Manganese ores, l. t............	10,000	23,258	30,118
Chromium or chrome-ore, l. t.....	2,000	2,700	2,000
Copper, s. t	71,186	82,938	78,368
Lead, s. t.....................	139,897	129,412	135,629
Zinc (metallic), s. t............	38,544	40,688	42,641
Zinc used as oxide (paint), s. t....	13,000	15,000	18,000
Nickel (metallic), lbs............	64,550	245,504	182,345
Tin (black tin) ore, s. t.........	....	200	200
Cobalt, lbs....................	2,000	8,423	8,689
Platinum, oz. troy..............	150	250	50
Iridosmine (used in pointing pens), oz. troy.....................	....	300	300
Antimony (metallic), lbs.........	61,208	(ore) 243,635	70,000
Aluminum, oz. troy	1,800	3,400	5,000
PRECIOUS METALS.			
Gold, est. by the mint	$30,800,000	$31,801,000	$35,000,000
Silver, est. by the mint..........	$48,800,000	$51,600,000	$51,000,000
Quicksilver, flasks (76½ lbs.)......	31,915	32,073	30,068
MINERALS (FUEL).			
Anthracite coal, l. t............	33,175,756	34,228,548	34,853,077
Bituminous, all varieties, l t......	73,730,539	64,840,668	63,178,248
Coke, s. t.....................	4,873,805	5,106,696	6,835,068
Petroleum, bar. (42 gals. each)....	24,089,758	21,842,041	28,110,115
Natural gas (value of coal displaced).....................	$1,460,000	$4,854,200	$9,847,150
MISCELLANEOUS.			
Salt, s. t.....................	912,091	985,368	1,078,991
Bromine, lbs	281,000	310,000	428,334
Borax (concentrated), s. t........	3,500	4,000	4,887
Sulphur, s. t..................	500	700	2,500
Pyrites, l. t...................	35,000	49,000	55,000
Barytes, s. t..................	2,500	15,000	10,000

	1884.	1885.	1886.
MISCELLANEOUS.			
Mica, lbs.....................	147,410	92,000	41,000
Feldspar, l. t..................	10,900	13,600	14,900
Asbestus, s. t.................	1,000	300	200
Graphite, lbs. (1883)......575,000	.....	327,883	415,525
Asphaltum, l. t................	3,000	3,000	3,500
Alum, s. t.....................	19,000	Not report'd	33,750
Ochre, l. t....................	7,000	3,950	14,542
Copperas, s. t	7,750	Not report'd	11,000
Mineral waters (gallons sold).....	68,720,936	*9,148,401	*8,950,317
Precious stones (American), value.	$82,975	$69,900	$79,058
Gold-quartz used as ornaments....	......	$140,000	$40,000
STRUCTURAL MATERIALS.			
Building-stone, value est.........	$19,000,000	$19,000,000	$19,000,000
Lime, s. t.....................	3,700,000	4,000,000	4,250,000
Cement (Portland artificial), s. t...	20,000	30,000	30,000
Natural cement-rock, s. t.........	585,000	600,000	870,000
Brick and tiles, value est.........	$30,000,000	$35,000,000	$38,500,000
ABRASIVE MATERIALS.			
Buhrstones, value..............	$300,000	$100,000	$275,000
Grindstones, value	$570,000	$500,000	$250,000
Corundum and emery, s. t........	Not report'd	Not report'd	(ore) 645
Novaculite (whetstones), s. t......	....	265	580
Lithographic stone, s. t........ ..	....	....	50
FERTILIZERS.			
Phosphate-rock (washed), l. t.....	431,779	437,856	430,549
Fertilizers manufactured, s. t.....	967,000	1,023,500	Est1,023,500
Gypsum, land plaster (native stone), s. t...................	60,299	75,100	50,000
Gypsum, calcined or stucco, s. t...	26,880	26,440	26,000

* Artesian wells excluded.

THE END.

The Human Species.

By A. DE QUATREFAGES, Professor of Anthropology in the Museum of Nat
ral History, Paris. 12mo. Cloth, $2.00.

The work treats of the unity, origin, antiquity, and original localization of t
human species, peopling of the globe, acclimatization, primitive man, formation
the human races, fossil human races, present human races, and the physical a
psychological characters of mankind.

Students' Text-book of Color; or, Modern Chromatics

With Applications to Art and Industry. With one hundred and thir
Original Illustrations, and Frontispiece in Colors. By OGDEN N. Roo
Professor of Physics in Columbia College. 12mo. Cloth, $2.00.

" In this interesting book Professor Rood, who, as a distinguished Professor
Physics in Columbia College, United States, must be accepted as a competent a
thority on the branch of science of which he treats, deals briefly and succinctly wi
what may be termed the scientific *rationale* of his subject. But the chief value
his work is to be attributed to the fact that he is himself an accomplished artist
well as an authoritative expounder of science."—*Edinburgh Review, October*, 187
in an article on "*The Philosophy of Color.*"

Education as a Science.

By ALEXANDER BAIN, LL. D. 12mo. Cloth, $1.75.

" This work must be pronounced the most remarkable discussion of education
problems which has been published in our day. We do not hesitate to bespeak 1
it the widest circulation and the most earnest attention. It should be in the han
of every school-teacher and friend of education throughout the land."—*New Yo*
Sun.

A History of the Growth of the Steam-Engine.

By ROBERT H. THURSTON, A. M., C. E., Professor of Mechanical Engineeri
in the Stevens Institute of Technology, Hoboken, N. J., etc. With o
hundred and sixty-three Illustrations, including fifteen Portraits. 12
Cloth, $2.50.

" Professor Thurston almost exhausts his subject; details of mechanism are f
lowed by interesting biographies of the more important inventors. If, as is c
tended, the steam-engine is the most important physical agent in civilizing 1
world, its history is a desideratum, and the readers of the present work will ag
that it could have a no more amusing and intelligent historian than our author.'
Boston Gazette.

Studies in Spectrum Analysis.

By J. NORMAN LOCKYER, F. R. S., Corespondent of the Institute of Fran
etc. With sixty Illustrations. 12mo. Cloth, $2 50.

" The study of spectrum analysis is one fraught with a peculiar fascination, a
some of the author's experiments are exceedingly picturesque in their results. Th
are so lucidly described, too, that the reader keeps on, from page to page, ne
flagging in interest in the matter before him, nor putting down the book until 1
last page is reached."—*New York Evening Express.*

New York: D. APPLETON & CO., 1, 3, & 5 Bond Street.

eneral Physiology of Muscles and Nerves.

By Dr. I. ROSENTHAL, Professor of Physiology at the University of Erlangen. With seventy-five Woodcuts. ("International Scientific Series.") 12mo. Cloth, $1.50.

"The attempt at a connected account of the general physiology of muscles and erves is, as far as I know, the first of its kind. The general data for this branch of cience have been gained only within the past thirty years."—*Extract from Preface.*

ight:

An Exposition of the Principles of Monocular and Binocular Vision. By JOSEPH LE CONTE, LL. D., author of "Elements of Geology"; "Religion and Science"; and Professor of Geology and Natural History in the University of California. With numerous Illustrations. 12mo. Cloth, $1.50.

"It is pleasant to find an American book which can rank with the very best of oreign works on this subject. Professor Le Conte has long been known as an original investigator in this department; all that he gives us is treated with a masterand."—*The Nation.*

nimal Life,

As affected by the Natural Conditions of Existence. By KARL SEMPER, Professor of the University of Würzburg. With two Maps and one hundred and six Woodcuts, and Index. 12mo. Cloth, $2.00.

"This is in many respects one of the most interesting contributions to zoölogical iterature which has appeared for some time."—*Nature.*

he Atomic Theory.

By AD. WURTZ, Membre de l'Institut; Doyen Honoraire de la Faculté de Médecine; Professeur à la Faculté des Sciences de Paris. Translated by E. CLEMINSHAW, M. A., F. C. S., F. I. C., Assistant Master at Sherborne School. 12mo. Cloth, $1.50.

"There was need for a book like this, which discusses the atomic theory both n its historic evolution and in its present form. And perhaps no man of this age ould have been selected so able to perform the task in a masterly way as the illusrious French chemist, Adolph Wurtz. It is impossible to convey to the reader, in notice like this, any adequate idea of the scope, lucid instructiveness, and scienific interest of Professor Wurtz's book. The modern problems of chemistry, which re commonly so obscure from imperfect exposition, are here made wonderfully clear nd attractive."—*The Popular Science Monthly.*

he Crayfish.

An Introduction to the Study of Zoölogy. By Professor T. H. HUXLEY, F. R. S. With eighty-two Illustrations. 12mo. Cloth, $1.75.

"Whoever will follow these pages, crayfish in hand, and will try to verify for imself the statements which they contain, will find himself brought face to face vith all the great zoölogical questions which excite so lively an interest at the present day."

"The reader of this valuable monograph will lay it down with a feeling of wonler at the amount and variety of matter which has been got out of so seemingly light and unpretending a subject."—*Saturday Review.*

Suicide:

AN ESSAY IN COMPARATIVE MORAL STATISTICS. By HEN MORSELLI, Professor of Psychological Medicine in the Royal Universi· Turin. 12mo. Cloth, $1.75.

"Suicide" is a scientific inquiry, on the basis of the statistical method, into the la of suicidal phenomena. Dealing with the subject as a branch of social science, it c siders the increase of suicides in different countries, and the comparison of nations, rac and periods in its manifestation. The influences of age, sex, constitution, climate, seast occupation, religion, prevailing ideas, the elements of character, the tendencies of civili tion, are comprehensively analyzed in their bearing upon the propensity to self-destr tion. Professor Morselli is an eminent European authority on this subject. It is acco panied by colored maps illustrating pictorially the results of statistical inquiries.

Volcanoes:

WHAT THEY ARE AND WHAT THEY TEACH. By J. W. Jui Professor of Geology in the Royal School of Mines (London). Wi Ninety-six Illustrations. 12mo. Cloth, $2.00.

"In no field has modern research been more fruitful than in that of which Profes Judd gives a popular account in the present volume. The great lines of dynamical, g logical, and meteorological inquiry converge upon the grand problem of the interior c stitution of the earth, and the vast influence of subterranean agencies. . . . His bool very far from being a mere dry description of volcanoes and their eruptions ; it is rat a presentation of the terrestrial facts and laws with which volcanic phenomena are as ciated."—*Popular Science Monthly.*

"The volume before us is one of the pleasantest science manuals we have read 1 some time."—*Athenæum.*

"Mr. Judd's summary is so full and so concise, that it is almost impossible to give fair idea in a short review."—*Pall Mall Gazette.*   ·

The Sun.

By C. A. YOUNG, Ph. D., LL. D., Professor of Astronomy in the Colle of New Jersey. With numerous Illustrations. Third edition, revise with Supplementary Note. 12mo. Cloth, $2 00.

The "Supplementary Note" gives important developments in solar astronomy sin the publication of the second edition in 1882.

"It would take a cyclopædia to represent all that has been done toward clearing up t solar mysteries. Professor Young has summarized the information, and presented it a form completely available for general readers There is no rhetoric in his book ; truets the grandeur of his theme to kindle interest and impress the feelings. His sta ments are plain, direct, clear, and condensed, though ample enough for his purpose, a the substance of what is generally wanted will be found accurately given in his pages.' *Popular Science Monthly*

Illusions:

A PSYCHOLOGICAL STUDY. By JAMES SULLY, author of "Sensati and Intuition," etc. 12mo. Cloth, $1.50.

"An interesting contribution by Mr James Sully to the study of mental patholo; The author's field of inquiry covers all the phenomena of illusion observed in sense-p ception, in the introspection of the mind's own feelings, in the reading of others' feelin in memory and in belief. The author's conclusions are often illustrated by concrete e ample or anecdote. and his general treatment of the subject, while essentially scientific, sufficiently clear and animated to attract the general reader."—*New York Sun.*

New York: D. APPLETON & CO , 1, 3, & 5 Bond Street.